THAT WHICH ROOTS US

THAT WHICH ROOTS US

THAT WHICH ROOTS US

Environmental Issues in the Pacific Northwest and Beyond

MARION DRESNER

UNIVERSITY OF NEVADA PRESS | *Reno & Las Vegas*

University of Nevada Press | Reno, Nevada 89557 USA
www.unpress.nevada.edu

FIRST PRINTING

Cover design by Louise OFarrell
Cover photograph by Amy Reddick Wilkes

Library of Congress Cataloging-in-Publication Data on file
ISBN 978-1-64779-112-4 (paper)
ISBN 978-1-64779-113-1 (ebook)
LCCN 2022062217

The paper used in this book meets the requirements of American National Standard for
Information Sciences—Permanence of Paper for Printed Library Materials, ANSI/NISO
z39.48–1992 (R2002).

All royalties coming to the author will be donated to organizations fighting climate change.

Contents

Contents

THAT WHICH ROOTS US

THAT WHICH ROOTS US

Introduction

GLOBAL CLIMATE CHANGE and serious loss of biodiversity are two of the most pressing problems of our time. Most of us are out of touch with planetary limits and ecological constraints, yet these crises demand us to change. Motivation to help restore ecosystems and to mitigate climate change comes out of caring and feelings of connection to the more-than-human world. By spending more time outdoors, observing trees, birds, and animals living nearby, we can learn to sense them as being as vitally alive as we are.

The book begins with a walk in a giant redwood forest. Redwood National Park came into existence because of the dedicated work of scientists and environmental activists. Although groves of redwoods are now preserved, most of our forests are only shadowy relics of past grandeur. A century of clear-cutting and fire suppression, growing drought and summer heat events caused by climate change are the culprits. I describe some of the restoration efforts and action to preserve existing forests.

Many people continue to behave as if the not-human world is inert. We might be inspired by the traditions of Indigenous peoples in America, who have not lost their vibrant sense of mutuality with that world. The predecessor of modern Europeans, Cro-Magnon, took the time to reflect on life's magnificence, having vital relationships with nature. The evocative paintings inside Paleolithic caves are hopeful—there are no cave paintings depicting war, avarice, or greed. Cro-Magnons seem to have lived as equals; there are no signs of a hero or a king. The paintings of animals reflect respect, reverence, and reciprocity. A reverence for wild animals was supplanted by a reverence for human ancestors as farming supplanted hunting and gathering. The chapter describes some modern agroecological methods that recognize the value for soil organisms and insects and animals that help keep a farm productive without causing further harm.

Western sympathy with wild nature reawakened in the age of the English poet William Wordsworth. He encouraged others to become aware of the outer wild and their own inner wildness. He inspired the first public parks in England for ordinary people to visit. His writings inspired American transcendentalists Emerson and Thoreau. John Muir was inspired by Thoreau, and America's national parks were inspired by John Muir's writings. A visit to an intertidal zone prompts thoughts about other heroes of conservation, including Rachel Carson.

Oregon's beaches and coastline were saved from overdevelopment by the foresight of then-governor Tom McCall over fifty years ago. Giant remnant old-growth Sitka spruce forests and massive geologic forces of the past are evident all along the coastline. Restoration efforts along the Oregon coast include the return to prescribed burning and the return of sea otters through partnerships with the tribes and environmental advocates.

Some of the many avenues for experiencing the world with a sense of awe and concern are described: kayaking with orcas, protecting bumblebees, working for greater equality, helping wild birds by renaturalizing backyards in the city. Wild species can rebound if we provide more habitat and protection from toxins.

There are indications that we are coming to understand our ecological limits. Dams, once a potent symbol of the power to "harness nature," are being decommissioned. Removal of dams, like on the Klamath River, brings healing for an entire river ecosystem. Indigenous people collaborate with Western scientists to conduct prescribed burns and mitigate the damage from future fires. Forest bathing, walking in a forest while trying to consciously connect with life around you, is a popular activity. Regenerative agriculture is increasingly recognized as essential to restoring soil health and sequestering more carbon. A growing number of people are concerned about global climate change and are acting politically to help slow it.

When we see ourselves as interconnected with other beings, we begin to understand ecology and global climate change, and we begin small acts of caring for the living world. Beginning in childhood, there are many opportunities for transformations in our understanding, to build an emotional affinity with nature through the simple act of paying attention.

I hope this book can help you, the reader, pay more attention to the intricate ecological connections in the world. Try to root yourself to the living processes and beings around you and identify how you might help bring about the changes we need.

CHAPTER 1

The Redwoods

THE NATURAL WORLD is in trouble, and so are we. We depend utterly on the tangled net of ecological relationships, but they are breaking apart because of species extinctions. If we try to understand the lives of the trees, the birds, and the animals living near us, if we see them as alive as we are ourselves, will we act with more empathy? If we understand ourselves as part of the community of nature, will we act more responsibly? With these questions in mind, I take a series of journeys to natural places significant to me. During these journeys and in reconstructing the story of where humans come from, I see the vibrancy of the natural world.

Culture shapes our identities and values through stories, especially ones that tell us where we came from. Living alongside animals who brought meaning to their lives, Ice Age Europeans probably understood the interconnections of the world. Many Indigenous people still understand their kinships with all life. Animals are symbolically represented in origin stories. But most people living in modern Western culture are cut off from daily life lived among other species. While many try to rediscover a reverence for nature, the dominant culture eats away at the possibilities for reconciliation.[1] Wondering what one person can do, I head for a long walk in the forest.

I adored the giant sugar maples shading the street where I grew up in New York City. Before I had ever seen a redwood tree, I cherished them too. I was enthralled when I saw a photo of a grove of ancient redwoods placed in a newspaper by the Sierra Club. I was a teenager when I read: "History will think it most strange that America could afford the moon...while a patch of primeval redwood trees was considered beyond its means. Save the Redwoods!"[2] I was gripped by the image of beautiful, stately redwoods—and appalled that they were rapidly being felled. Horrified and disgusted by what adults were doing to the world, I resolved to help. I sent $25 from money I had earned babysitting to the Sierra Club to save the redwoods.

I left grimy New York City eight years later. Hitchhiking to Northern California, I encountered wide-open places, vast beaches, forested mountain ranges, and the redwood forest.

FOREST ASSOCIATIONS

Years later, Roger and I camp on the banks of the Smith River among the ancient redwoods of Northern California. The world's tallest growing trees are all around us, their thick furrowed red bark absorbing noise and generating a deep quiet. As I walk among them, their grandeur seeps into me. And like it did for generations of philanthropists, scientists, lobbyists, and activists, their beauty motivates me to do the right thing.

We set out on our hike in Jedediah Smith Redwoods State Park. Crossing the river over the swaying wooden footbridge, admiring the clear blue-green water, we step over the mounded small river rocks on the other side and turn up onto the bank toward the trees. Soon we are walking alongside rippling Mill Creek. It's sunny and warm. Clouds of insects circle upward. Under the canopy of the redwoods, it's cool and moist enough for a variety of ferns to grow. Five-fingered fern, with leafy fronds looking like feathery hands, grows among clumps of tall sword fern and diminutive deer fern. As we continue, our footfall is softened by layers of redwood needles that cover the forest floor.

Roger and I climb higher along the trail and cross a small wooden bridge strengthened with triangular crossbeams on either side. Looking down at a tiny tributary that descends to Mill Creek, we admire shady pools and cascades. Delicate five-fingered fern grows alongside the pools. The trail twists around the deep dark green pools of Mill Creek. We enter a lost world of magnificent old-growth redwoods. I gaze up to the barely visible network of crowns held aloft by the big trees.

We head to El Viejo del Norte, the fifth-largest redwood, living in the Grove of Titans. As we walk I'm looking down to avoid tripping on roots when Roger nudges me—"Look up!"

To our left is a broad, incredibly tall redwood—our eyes scan upward along its massive reddish-gray fluted trunk. About fifty feet up, a huge bulky elbow of a burl protrudes out and then upward. Nearby stand more giants: the Lost Monarch, the Screaming Titans, El Viejo del Norte, and others. I stop, breathing slowly so as to better take in the details of their majesty. As my eyes scan them from the crown down to the forest floor, and up again, the thought of what might have happened to these beautiful giants wrenches me. These were saved from destruction. But I wondered which other formerly magnificent trees were lost, having been turned into a commodity.

What lives up there in the canopy, unseen from the ground? What I can see from the ground are fuzzy-looking puffs of foliage surrounding topmost branches. These canopies are physically complex, say the researchers who climb 150 feet up into the redwoods. The crown of a single ancient tree may look like an entire forest stand. After enduring hundreds of years of wind storms and lighting strikes, a big tree's branches have broken off. This increases light high up on the trunk, which coaxes the dormant bud tissue up there to sprout. The sprouts may grow as erect as other trunks growing on the ground. Each resulting trunk can grow sturdy enough to support its own branches, and the trunks and branches of these can fuse to each other in a process called reiteration. The Del Norte Titan has forty-three different branching trunks; one grouping of canopy trunks collected soil three feet deep in a hollow high above the forest floor. The tree even sends roots into it to drink the moisture and nutrients. Canopy researchers describe and catalogue plants living in this aerial forest: evergreen and red huckleberry, wads of leather fern, mosses, rhododendron bushes in bloom, fruiting red currant and elderberry bushes, bonsai California laurel, western hemlock, Sitka spruce, Douglas fir, and small redwood trees growing upward from the massive interlocking branches. The canopy houses tiny mites, beetles, earthworms, millipedes, bumblebees, egg masses from breeding clouded salamanders, red tree voles, and nesting marbled murrelets.

Murrelets, dove-sized seabirds, are rare in the Pacific Northwest. Whereas other seabirds like puffins nest in colonies on offshore rocks, murrelets nest in the high, broad branches of old-growth trees, selecting a mossy platform to lay and incubate a single egg. Each morning, the parents fly out to sea to forage for small fish, like herring, returning in the dark of evening to avoid predators that might find their nest. Logging the old growth reveals their nests, making them more susceptible to predation. Laying just one egg a year, they recover slowly from a disturbance.

Around the bases of many redwoods, tall trunks grow from bud tissue in the same way the trunks in the reiterated crown grow. Continuing, we reach a massive redwood lying on its side, so wide it towers over us even in decay. We walk through a tunnel made by the overhanging vine maple branches that reach over the enormous trunk, dripping with moss and lichen. As it decays, the log adds coarse woody debris to the forest floor, helping it hold onto moisture during the long summer drought. The trail turns closer to the creek. Below us, the creek forms a deep, dark green pool of water, an inviting reprieve from the heat. We descend, strip off our clothes on the sandy bank, and dive in, laughing and splashing each other. Afterward, we head back along Mill Creek to its confluence with the Smith River.

DEFENDERS OF THE FOREST

About a hundred years ago, these very redwoods were threatened with logging. Future-minded wealthy conservationists bought up groves of redwoods and formed the Save the Redwoods League. Ironically, many of these conservationists used inherited money obtained by plundering the planet elsewhere. Some were racist, considering those of white "races" more desirable than people of color. Joining the protection efforts were women empowered by the growing feminist movement. In the 1920s, groups of women gathered to protest logging of the ancient redwoods. Wearing broad hats and long skirts, they surrounded giant trees to defend them from loggers. These forest defenders foreshadowed tree sitters who protected the forest sixty-five years later.[3]

By the 1940s, many groves along the Smith River were stitched together to create Jedediah Smith Redwoods State Park. But too much ancient forest was still held by private timber companies. A postwar building boom in the 1940s expanded logging on these redwood lands. New groups of concerned preservationists urged politicians to form a new national park. Logging companies fought back, insisting that old growth was senescent, like a grove of senior citizens dozing in a retirement home.

Then, the *National Geographic* magazine published remarkable photos of the magnificent Tall Trees Grove along Redwood Creek, appealingly naming them the "Mount Everest of all living things."[4] Public attention was focused on these trees, and events moved rapidly. A call for a Redwood National Park began. Logging companies countered this appeal, warning that creating a park would cut local employment. The Sierra Club published that famous *New York Times* ad that had caught me up as a teenager. Lucille Vinyard, the local Sierra Club leader, led hikes and float trips, promoted the park at talks, took photos, and testified at government hearings. Finally, in October 1968, after years of advocacy and diplomacy by citizen activists, the new national park was created. Existing redwood state parks were intertwined with new federally owned lands, protecting the Tall Trees Grove and other lands.

Redwood National Park's boundaries were not sustainable. They were not created using ecological realities, but by political compromise between the government and logging companies. In one crucial watershed, the boundaries stretched from the lower reaches of Redwood Creek along a thin ribbon of old growth, extending upstream to include the now legendary Tall Trees. Upslope and upstream from the boundary, two-thirds of the watershed spread. Realizing their access to these lands would soon end, the timber companies cut every tree on the slopes, extracting over a billion board feet of redwood. The bare hillslopes could not hold onto the rainwater. Water flowed overland down the slopes, washing soil, huge plugs of rock, and logging debris downslope, eventually tumbling into Redwood Creek. This debris raised the streambed's level as much as fifteen feet in some places.[5] The big redwoods growing along the stream-banks were in jeopardy. Redwood roots are shallow, roots can drown without the necessary oxygen, and a redwood can topple over during a flood event. Forest ecologists were anxious not to have a replay of the Christmas floods of 1964, when hundreds of old redwoods fell in Bull Creek in Humboldt Redwoods State Park.[6] Upstream clear-cutting had exposed bare soil on steep slopes to torrential rains. Soil and rock washed downslope and downstream. The flooded stream overflowed its banks, knocking over giant trees in Humboldt Redwoods, undercutting stream banks, toppling other giant trees, and suffocating the roots of even more.

Renewed efforts to enlarge the park's boundaries began in earnest, reigniting bitter conflicts between logging companies and conservationists. Vinyard, the Sierra Club activist, was threatened with a knife. Another Sierra Club activist—Dave Van de Mark—was bullied by thugs who warned, "We know where you live."[7] Meanwhile, logging continued on the steep slopes above the park boundary.

The national park boundaries were finally extended. This expansion set a legal precedent. Land management with other landowners in the Redwood Creek watershed became more cooperative over time. In other places, boundary lands adjacent to other national parks began to be managed more cooperatively with the National Park Service.[8]

MARY ANN MADEJ and her family often camp alongside our family during our annual visit to the redwoods. She was hired to help rehabilitate the lands damaged by prior logging when the park was expanded. She and the other young, energetic geologists, hydrologists, and biologists of that time had strategized how to best proceed. Still, on her first tour of the area, she was shocked to see the barren hillsides, which once supported old-growth forest. Where the land had not slid into the creek, the redwood stumps stood like gravestones. In the riverbed, vast expanses of sterile gravel stretched from bank to bank. Where once there had been deep pools for habitat for steelhead and salmon, now there were none. Redwood Creek was filled with twenty-five feet of excess gravel and sand in many places. The heightened river cut into its own banks, causing further landslides and stream-bank erosion. Streamside redwoods were dead or dying from this excess sedimentation.

As Mary Ann's team proceeded, they observed how active landslides and gullies that poured sediment into streams and rivers during each winter storm originated at abandoned logging roads and logged hillslopes. These logging roads were the main culprit contributing to erosion and increased sediment; so Mary Ann and her colleagues began the work of many years, removing hundreds of miles of logging roads. At first they built water bars, check dams, and other labor-intensive erosion control features, mostly by hand. Then they realized that the same large heavy equipment that was used to construct roads would be more efficient to remove them. By the 1990s, each road bench was completely recontoured with material excavated from buried stream channels and restored to their former, natural shape. Then the hillslopes were replanted. Lessons learned from those efforts have since been applied to other forested lands nationwide; even landowners of privately owned timberlands now use road restoration techniques.

To measure the effectiveness of road restoration, crews monitored the recovery of Redwood Creek over several decades. This involved monitoring streamflow and sediment during storms, both at night and during the day. During one storm, Mary Ann was dipping stream sampling equipment into floodwaters while dangling from an open metal cart suspended by a cable above the swiftly moving river while branches crashed down around her. A young redwood toppled across the cableway, preventing Mary Ann and her co-worker, Greg, from getting back to their truck. They bushwhacked six miles up steep hills in the rain, continuing in the dark until they reached a gravel country road, where they hitched a ride back to town.

Mary Ann and her colleagues also surveyed pool depths and snorkeled through each pool to count fish during the summer. They also collected and examined aquatic macroinvertebrates. Using 1975 data as a benchmark, the team discovered stream life diversity in the creek had significantly improved over time. As Redwood Creek cut through the thick sand and gravel deposits, it began flushing excess sediment downstream. Less and less fine sediment was added from upslope into the stream channel. Pools for salmon are now more frequent and deeper, and material in the channel bed changed to gravel. Salmon prefer gravel over finer material to lay their eggs. All of these changes have improved salmon habitat. Efforts to better stream habitat for Coho and Chinook salmon continue today. The river system recovered, showing its resilience.

Redwood National Park contains once logged over lands, with small, young trees. The park's mission is to provide an unimpaired redwood forest for the future, so most ecologists want to help the process along by restoring these younger groves. On another day, we meet NPS forester Jason Teraoka, a tall, friendly man sporting a long ponytail of thick black hair, and we head upslope. Jason explains that most of the restoration work is in upland redwood forests, because that is where the logging had occurred.

Jason shows us some features of a healthy redwood forest. With trees ranging from two hundred years old to much younger, the forest sports multiple vertical layers of branches and needles. There are many snags, or standing dead trees, and lots of decaying wood on the ground. A diverse combination of shrubs, ferns, and flowering plants grows on the forest floor in the light that penetrates there.

Jason drives us over to a patch of forest that was clear-cut fifty years ago, burnt and reseeded by the logging company. The forest is tall and skinny, with densely packed redwood sprouts, Douglas fir,

Sitka spruce, and western hemlock trees, growing as if in a race to reach the top for sunlight. The canopy is invisible high above us. Jason explains that the trees are so tall and leggy because they have put their energy into height and don't have enough needles in relation to their height. No light hits the ground here, so nothing grows on the forest floor. There is no downed and decaying wood.

The goal of thinning is to mimic the tree density found in old-growth forests, to provide light and space so remaining trees can grow taller and wider. Remaining trees will grow larger and may resemble old growth over a shorter time span. To observe this for ourselves, we travel to a stand that Jason's team thinned ten years ago. Here, light penetrates down to the forest floor through gaps created from the thinning, allowing native ferns, wildflowers, and shrubs to grow. Looking up, I can see the leafy canopy extending much lower down the trunks. Jason has obvious pride in how well the stand has recovered, becoming a spacious forest of vigorous redwood, Douglas fir, and western hemlock. Downed decaying wood was left behind to provide moisture holding and habitat for insects and small animals. "If you create the structure, they will come," Jason says. "If we did nothing until trees topple by their own accord, it would take centuries for natural processes to create breaks in the canopy and allow sunlight to penetrate." He explains how amphibians, birds, and small mammals not here before are now using this forest stand, and over a much shorter time frame than if foresters had left the stand on its own.

Continuing up the road to Holter Ridge, we pass a logging truck carrying a load of small-diameter logs. "Are they logging upstream?" I ask when we stop and get out to view the forest. He tells us: "Those are 'our' guys." He explains that the park now collaborates with local businesses, many whom were once enemies of conservation, to do the thinning. They allow the contractors to extract some merchantable trees to provide incentive for the loggers to pull out the old logging roads at no cost to the Park Service. They lop the rest of the debris and scatter it throughout the forest to make a layer of coarse downed woody debris.

Allowing logging in a national park is controversial. Logging the redwoods especially makes environmentalists angry. "Fifty percent of my job has been PR," Jason tells us. "Trying to work with environmental groups, the state and national park personnel, and local lumber companies to make this work." But I can see for myself how the trees now have an uneven, multi-aged look. Saplings grow up from the forest floor where thinning has occurred.

Jason leaves us at the top of a watershed called Lost Man Creek. Downstream from us, in the old-growth forest, the creek continues to flow past ancient redwoods along the stream valley, among five-fingered fern, wild ginger, and sword fern.

REDWOODS are the tallest trees now existing—they achieve this by growing in height for their first hundred years. But how does a tree that grows so tall get water from its roots up to its needles? The trip from roots to the crown takes water over two weeks. According to tree physiologists, the water actually gets pulled up by a combination of factors. Differences in pressure between soil and the thin-walled root cells, coated with mycorrhizae, help move water from the soil into the roots, but these forces can account for only a tiny amount of pull. The force that moves water up may also be driven by the affinity water molecules have for each other—picture water molecules forming a long continuous column from outside the root up to the leaves or needles. Once inside the root, water moves through a system of long and narrow interconnected cells called xylem—only thirty microns wide, or half the diameter of a human hair—in the wood and in the branches. Inside the xylem are stacked cells with tiny holes called tracheids that transport water; the cells are stacked one on top of another like a living system of pipes. Water moves upward from one tracheid to another. As water evaporates from the leaves through nanoscale pores, one molecule escaping into the atmosphere from a redwood needle pulls another molecule behind it in a continuous process. Amazingly, most of the water a tree absorbs—95 percent—moves up and out into the atmosphere. It is not retained by the tree for photosynthesis or for building new cells.[9] The big trees actually take water vapor back to the lower atmosphere.

Redwoods are living elders that give us an alternative relationship to time. They're known to live as long as 2,200 years. What was the riverbank like when these trees sprouted? What conditions enabled them to reach this size? Will giant redwoods exist in the future? Climate change will increase storms and flooding in the Pacific Northwest, but we also expect severe droughts and wildfire. So much of the redwood range is not old growth; only a tiny percentage of ancient forests remains. Can these younger trees ever become future groves of ancient trees?

Fog in the coastal redwood forests of California delivers a sizable portion of water during the long summer drought, helping relieve summer water stress and sustaining growth. Only old-growth redwood stands, with their complex canopies, can capture the fog during the long summer drought. Fog drip may be the necessary ingredient that allows redwoods to reach their astounding heights. Fog drip makes it to the forest floor, providing almost all of the water needs in the summer months for other plant species.

Fog along the coast of Northern California has been decreasing; during the fog season there is three hours' worth less per day than previously.[10] Redwood trees may become increasingly water stressed. This could mean trouble for stands of younger, regenerating trees, which may not ever catch up.

CHANGING HOW FORESTS FUNCTION

The forests of North America once seemed to stretch out endlessly, fostering a mentality that forests were endlessly renewable. About one-half of the United States was covered in old-growth forest when the first European explorers arrived. A traveler in Ohio in the mid-1700s saw thousands of oaks "14–15 feet in circumference, their straight stems rising...to 70 or 80 feet."[11] Vast, pristine swaths of forests covered much of Michigan, Wisconsin, and Minnesota. Trees growing along the West Coast were enormous. Charles Wilkes sailed along the coast of Washington in 1840 for the U.S. Exploring Expedition. Their route passed the giant ancient cedar forest, where the trees were the diameter of a grown man and stretched over two hundred feet high. Wilkes remarked, "I could not control my astonishment."[12]

But what comes back, after being over-exploited, isn't the same as what was once there. The magnificent forests the early settlers saw will probably never exist again. Europeans who had discovered a "new world" began to exploit it as soon as they arrived in North America. Euro-Americans arrived in North America without a tradition of understanding ecological limits. Not caring about the consequences, they substantially altered western North America's forests through drastic cutting and by suppressing fire.

Before the period of fire suppression and intensive logging of large trees, there were many big fire-tolerant old trees in many watersheds. These large stands of fire-resistant old-growth trees gave a natural resistance to wildfires and other disturbances. Having thick bark that insulates the tree is an adaption to survive fire. Mature ponderosa pine trees, the dominant tree of the eastern Cascades, also have long needles in tight bunches that resist combustion, because they are elevated and therefore away from a typical flame zone.[13] When low-intensity fires are allowed to burn, trees get spaced far apart from one another in a parklike structure the early pioneers witnessed when settling in the area. On the west side of the Cascades, mature Douglas fir are tall, with branches and needles usually in the upper half of the tree, making it unlikely that the canopy will flame in typical fire conditions. Older Douglas fir have thick bark and extensive root systems that add to their fire resistance. Forest ecologists studied burn scars in old-growth forests and discovered that many old-growth trees had persisted through fires.[14] Fire and wind helped create different patches of vegetation types and mosaics of different stages of succession over time. It is this mosaic pattern of tree ages and species that provides the overall forest's resilience to disturbances like wildfires. But commercial tree plantations, created after clear cuts, grow in homogeneous patterns, easier for fire to move through and become destructive.

Fire was a land management tool commonly used by Indigenous peoples in the Pacific Northwest as part of their cultural traditions. As a result, western forests were once more tolerant of fire. But the government forbade "Indian" burning, beginning in Oregon in the 1850s and then in other western states. The forests subsequently became overgrown.

Naturally occurring forest fires have been actively suppressed since the Great Fire of 1910. Called the Big Burn, this firestorm burned over three million acres of the Northwest. More than a hundred fires were sparked by coal-burning trains traveling through forests made dry from drought. Then hurricane-force winds pushed the fires through western Montana and northern Idaho and northeast Washington to form one of the biggest fires in U.S. history.[15] Soon afterward, a widespread policy of suppression of any wildfire was enforced. Forests grew denser and more overgrown with small trees.

The backbone of large, old, fire-tolerant trees has been diminished by logging over time. Some of biggest trees were logged from private low-lying river valleys. Micha Ewers, who analyzed historical records, found references to giant 350-feet-tall Douglas fir trees that had stood in Washington State and British Columbia, but were felled before 1920.[16] Then, the logging companies' attention turned to forests on federally owned lands. Intensive logging of large, thick-barked, fire-tolerant ponderosa pine, western larch, and Douglas fir continued unabated until the Northwest Forest Plan of 1996. More than two-thirds of all the old-growth forests are gone. Young trees that are planted for eventual harvest in their place are much less fire resistant. Western forests have become less resilient, more susceptible to large crown fires and insect outbreaks, and less able to withstand disruption.

The fires of 2020 and 2021 were worse than the Great Fire of 1910. There were record-high temperatures during the summer, low humidity, poor spring rains, which made the western Cascade forests really dry. In 2020, a windstorm carried hot air from the high desert in the eastern part of Oregon westward over the Cascade Mountains. Lightning strikes ignited hundreds of forest fires, which raced down river canyons. More than 10.2 million acres of forest burned in 2020, 7 million in 2021. For those of us then living downwind of these forests, the skies were orange with smoke and the sun blotted out for weeks.[17]

Two different techniques, controlled burning or prescribed fire and manually thinning of the forest, can lessen the threat of a serious fire. Studies have been run on Douglas fir and Ponderosa pine forests in the West, comparing forest stands that had been thinned and stands that had been both thinned and burned with low-intensity fires with stands that were untreated. Both the thinned tree plots and plots that were thinned and burned in controlled settings were more resistant to low-intensity fires and bark beetles, and were less likely to break in wind events. Individual ponderosa pine and Douglas fir had more room and grew more than the trees in the stands that had not been treated.[18]

After forests are disturbed, they restore themselves through a process called succession. As Pacific Northwest conifer forests mature, the kinds, structure of, and sizes of trees slowly change over time. Douglas fir is the iconic tree of the Pacific Northwest. It grows on sites that were cleared by a disturbance, and the trees slowly change the physical environment as they grow larger by blocking more of the sun, helping the forest floor become moister and more suitable for western hemlock and western red cedar. Hemlock, with its delicate branches with small needles, and western red cedar, with foliage looking like green lace, grow in the shadow of Douglas firs. These are known as shade-tolerant species or late-successional species. If large Douglas firs are growing along with smaller shade-tolerant tree species in the understory, the stand is considered to be regenerating well. But forest succession is not as predictable as it once was, resulting in different structures than before.

MY STUDENTS AND I monitored research plots in Portland's 5,100-acre Forest Park over ten years. Our plots contained Douglas fir, big-leaf maple, and some western hemlock and western red cedar. Some of our plots were placed in old-growth stands, resplendent with many species of trees, an uneven, varied canopy, thick layers of soil, and abundant coarse woody debris. The young trees growing there were shade-tolerant species. The soil was topped with thick layers of decomposing needles and dark brown beneath.

The other research plots were closer to the city and had been logged when Portland was developing in the late 1800s. Every tree had been taken. In these plots, there were hardly any young trees and very few shade-tolerant hemlock and red cedar. Bigleaf maple trees with multiple trunks and numerous fifty- to eighty-year-old Douglas fir mainly grew there. There were many fewer trees overall in this part of the forest, half as many trees when compared to the old-growth stands and almost no coarse woody or organic debris on the ground. The soil was orange colored, with barely any topsoil or decomposing needles. After having been clear-cut, the bare slopes had eroded rapidly. Past winter storms had washed downstream the topsoil that had been there.

Up in the forest canopy of these plots, the tallest conifers had a damaged appearance. Instead of a variety of lichen species, there were only a few bleached, stunted, and necrotic microlichen, and the canopy branches were very skimpy. Air pollution was the cause. The damaged canopy has a reduced ability to hold water and host nesting birds, small voles, salamanders, and arthropods.

Earthworms not native to this region are common in Forest Park. Their tunnels and castings become obvious once you begin to notice them. Abundant and voracious, these earthworms consume whatever organic matter they can find on the forest floor. Back when the soil had been undisturbed, there were over a hundred native earthworm species. Where the soil is still undisturbed, the rare Oregon giant earthworm, growing to four feet in length, can be found, eating organic matter like mosses, decaying conifer needles, bits of wood, and grass seeds and stems.

Ecologists once assumed forest succession is predictable. But not so for these once-logged-over urban forest stands. These were clearly not going to become the old-growth hemlock and red cedar forests they had once been. They are damaged by long-term soil disturbance and the continuing influences of invasive species and air pollution. The changes we saw are persistent. Some factors regulating forest tree cover have shifted. Forest succession might no longer be predictable here and elsewhere.

In New England and the Upper Midwest forests, scientists have documented similar changes. Witness trees were used 150 years ago to mark the location of property boundaries. Using them now, scientists have made maps of early conditions to reimagine what the pre-settlement forests were like. The regrown New England forests are not as they once were. Once there had been towering hemlock and massive beech trees. They are now full of maple and poplar. Maple is capable of sprouting after damage and can survive in soils ranging from wetland to drier types.[19] The old-growth hemlock, maple, and birch forests of Minnesota have regrown with no hemlock or birch. White spruce and balsam fir grow there instead.[20] These forests contain mainly shade-intolerant or early successional trees.

Widespread logging is the main reason these forests have changed. Shade-tolerant trees like conifers were at a disadvantage after clear-cutting. In the Pacific Northwest, the forest composition shifted from giant western hemlock to Douglas fir. In New England, the forest shifted from hemlock to broad-leaved deciduous species.[21] Some call this process "biological homogenization" because instead of a mosaic of different forest types, the landscape has similar trees in most areas. The species of trees now growing in broad areas of these forests are without historical precedent.

European forests have regrown and expanded rapidly since the nineteenth century, mainly on former agricultural land. But the plant diversity is different when compared with forests older than two hundred years. Although old enough to have developed old-growth characteristics, these forests have not returned to their former state. This may be due to legacy effects from farming. In northeastern France, large areas that were deforested during the Roman occupation to be used for agriculture were abandoned and reverted back to forest after Rome fell. Scientists look at these forests to understand long-term change due to agriculture. Certain chemical and structural soil properties mark the ancient Roman fields. Now those former fields grow different types of plants than in areas not farmed by Romans. The researchers think the effects of ancient farming on forest biodiversity may be irreversible on a historical time scale.[22] Even land use from long ago can lead to extinction for plant species and the birds and other animals dependent on them.

To a large extent, ecosystems are resilient. The ability of an ecosystem to bounce back after a disturbance depends in part on the diversity of species present. The more diverse species in an ecosystem—the thinking goes—the more resilient. Pockets of old forest serve as refugia for the species needed by nearby disturbed areas to help them recover. Different species have unique abilities to respond to a disturbance encountered by the system. An alder tree growing in a very young stand of trees fixes nitrogen from the air to the soil, helping to supply nitrogen "fertilizer" to other plants in the young forest stand. A lichen—*Lobaria*—also fixes nitrogen. Growing in the crown of older Douglas fir, the lichen falls to the ground and supplies nitrogen to an old-growth stand. Diverse species help an intact and diverse forest to maintain itself over time.

Different species support each other below ground. We know much less about this unseen world. Underground, one-celled bacteria, fungi, and protozoa, nematodes and micro-arthropods, earthworms, beetles and other insects, along with plant roots, work together to maintain healthy forest soil. Small organisms eat by-products of the larger organisms. Larger animals eat the smaller ones. Together, these organisms decompose plant and animal waste, recycle nitrogen and make it available to plants. Some organisms help glom together soil particles, which helps water to infiltrate and get stored.

Vital underground fungal members of a forest community are lost when it is clear-cut. The fungi in an intact forest maintain the humidity levels needed. Numerous kinds of mycorrhizal fungi, the tree's fungal partner that inhabits the roots, usually connect the tree to the fungal partners in similar trees—and in some cases to other species. Dubbed the original worldwide web, these fungal partners exchange nutrients and information about impending insect attacks, enabling the trees to ward off bacteria and other fungi.[23] In these ways, they provide medicinal services and long-term stability to the forest. Botanists think mycorrhizal fungi helped plants to colonize land by developing associations early on with photosynthesizing plants. Their nutrient-gathering capacity enabled the first land plants to adapt to the dry conditions.[24]

Forest ecologist Suzanne Simard advocates saving patches of older legacy trees and their interconnecting networks so forests might withstand future stresses, helping to regenerate forests with a diversity of species and forest structures.[25] Simard, in collaboration with other researchers, determined that trees grow larger when connected to other trees via mycorrhizal networks. The mycorrhizae contain nodes and links between old, mature "hub" trees and younger trees, extending even to other species. Suzanne discovered how trees respond to one another via belowground signals, including distress signals sent via the miles of mycelium. They also collaborate by sharing sugar with each other.

Mycorrhizae supply trees with up to 40 percent of the nitrogen they receive, as much as 50 percent of the water they use. They also trade between 10 and 40 percent of their stored carbon and their phosphorus. One study of a grove of large Douglas fir included a large old tree linked to 47 other trees, and projected to be connected to 250 others. Simard nicknamed those trees "Mother" trees.[26] She urges retaining those larger trees and their robust mycorrhizal networks when an area is logged to improve the health and survival of the future forest to be planted or regenerated. Seedlings that were thus connected were 26 percent more likely to survive than those that were not.

Leaving patches of old growth has many benefits. Old-growth trees help maintain a forest's fire resilience, even as fire frequency increases. In western Oregon, Washington, and Northern California, old forests occupy a small portion of their historical extent because of widespread timber harvest, underscoring the significance of their unique structural features, including large, old trees and complex forest architecture, which provide habitat for threatened and endangered flora and fauna. In a recent study of all fires that burned between 2004 and

2014, only a quarter of all fires had burned in old-growth refuges. Garett Meigs, from Oregon State University, and his colleagues discovered that these old-growth patches had survived large fires because of their relative fire tolerance and because they grew in moist valley bottoms where cool air pools.[27]

Forests are living stores of carbon, keeping it from entering the atmosphere as CO_2. Old-growth forests accumulate large quantities of carbon over centuries, significantly more than younger trees. Globally, about 1 percent of living trees with large diameters are left, yet they store half of all the carbon stored in the world's forests.[28] Forests in Oregon and Washington currently store an estimated 2,100 million metric tons of carbon; they continue to increase their carbon storage at 0.3 percent per year.[29] By restricting timber harvests in some stands and lengthening the time between harvests on other stands, we will retain more old growth. An intact forest stores CO_2 in the trees and soil, preventing it from accumulating and trapping heat.[30]

Forest cover influences air temperature. Under the tree canopy itself, summer air can be cooler by as much as 10°C (18°F) in an extensive tree canopy when compared with that in a nearby city. Forest cools the air by evapotranspiration—when water changes from a liquid to a vapor, it absorbs energy or heat. Air in the city near a road is hotter because the asphalt and dark rooftops absorb and reradiate back more solar radiation. Adding to this, motor vehicles emit heat and buildings leak heat. Together, this creates a phenomenon called the urban heat island.[31] Leafy, tree-lined streets and greenspaces with trees throughout a city can help create more urban cooling.

CAN TREES BE RESPONSIVE to our presence? The Tolowa people, an Indigenous tribe living in the redwood region, believe that natural entities are endowed with spirits and power. Tolowa people hold their world renewal ceremonies in the redwoods each summer. This tradition goes back in time to the beginning, say the elders. They believe human behavior is governed by spiritually based rules; their dancing helps renew, balance, and reestablish positive relationships between humans and the earth.[32] The world renewal dance is one way their traditions are passed on to younger generations. Native Americans sustained themselves for thousands of years by practicing reciprocity within a natural community. Many elders from these communities understand how animals whom they live among know more than the people do. Although outwardly they may behave as Westerners do, using chainsaws or snowmobiles, many Indigenous people try to renew their cultures and teach other people how to move with the forces of the natural world, yielding to it. Tolowa tribal members now work with the Forest Service in Northern California to help manage the forest by using fire to restore the landscape. Members of the Yurok Tribe, just to the south, use prescribed burning on their traditional hazelnut-gathering areas, while Karuk tribal members, working with other groups, plan to restore a pattern of low-intensity burning to help restore resilience to the Klamath River Basin.

It is vital to rework our connections and obligations toward the natural world. All of our forests are now a shadowy relic of past grandeur. What precautions can Western cultures introduce against a human tendency of self-interest and greed? How can we remember to feel awe in the presence of giant redwoods? Awe fosters an awareness of connection and of human limitations. Giant redwood trees are endowed with names to help us understand their unique identities: the Del Norte Titan, Atlas, Lost Monarch, Paradox, John Muir, and New Hope. The vigilance of activists who saved ancient forests in the past is needed again.

ROGER AND I ARE CAMPED in the redwoods beside the Smith River again this summer. During the evening, we sit around the campfire with our daughter, her husband, and with friends of many years. Enjoying one another's company, we share stories.

Mary Ann tells us about a young girl who recently rescued a murrelet chick. The girl had attended a campfire program about marbled murrelets, where she saw a photo of a brown and white murrelet chick sitting on a high redwood branch. The next day, she found a young bird struggling on the ground. She recognized the bird. She ran over to the ranger station to tell them about the bird fluttering on the ground. Soon, a team was dispatched in kayaks to bring the fledgling out to the offshore rocks. The girl was in one kayak, helping return the bird to its habitat. We all loved this rescue story, feeling gratified that the bird was helped, hoping the girl who helped the bird would continue helping the wild.

Early the next morning I walk the dogs over to the river near where our tent is pitched. Mist sits in the redwood treetops; I imagine how the drip from the fog is giving the redwoods a much-needed drink. The entire watershed of this river is mostly forested. A splashing downstream. I turn my gaze out toward the sound. Several small, dark animal heads emerge, then quickly disappear beneath the surface. As they dive, their webbed feet show above the waterline. It is a family of river otters. They dive, surface, and dive again. Here in the Smith River, the last free-flowing river in the state, they have plenty of trout and even young salmon to catch. California had enough rain last winter to take it out of a drought, and the otters show us that, for the present, all is well. Delighted, I continue watching them until they swim away and out of our view.

CHAPTER 2

The Painted Caves

DURING THE MOST RECENT ICE AGE, herds of mammoth, bison, and reindeer thundered across the steppes. The coastline contained rich estuaries teeming with fish, and whales swam in the sea. Ice-Age people left records of their powerful and mysterious relationships with these animals in paintings made in caves and engraved on bones. In Europe, they were called Cro-Magnon after the place in France where their remains were first identified. Today we can see images of magnificent animals—painted or carved—deep inside limestone caves for ourselves.

Indigenous people around the world continue to follow traditions of reverence toward animals. For many traditional people, the world is alive—the animals and plants, and the rocks, mountains, and rivers. Cultures as diverse as the Koyukon people in Alaska and Aboriginal Australians believe an underlying pattern connects people with all the other beings of the world.

We visited an Oregon rock shelter and then traveled to France to experience Ice Age art for ourselves.

CASCADIA CAVE

The Cascadia Cave rock shelter sits on a hill above the South Santiam River near Salem, Oregon. It is where the Kalapuya inscribed animals sacred to them in rock art. Inside, on a rock wall, massive zigzag lines run horizontally across the cave. Rows of vertical lines, human faces made of holes, symbols representing male and female genitalia, feathers, salmon, concentric circles, and points of light are inscribed. A row of unmistakable bear tracks, surrounded by red ochre, is inscribed over the long zigzag line. Below that is a long horizontal crack in the wall. According to anthropologist David Lewis, these petroglyphs indicate this as a place of Bear power.[1] Bears were seen as humanlike in their ability to locate and dig for roots to self-medicate. Elders say their ancestors were taught by the bears which plants are good for medicine and which are good for food.

Cascadia Cave contains features and inscriptions indicating where a tribal shaman might enter the supernatural world, says archaeologist Tony Farque. Ancestors were available for consultation with a shaman about how to heal and how to live. In Cascadia Cave, symbols were painted—concentric circles, grid marks, and zigzag lines—that indicate entoptic phenomenon experienced during altered states of consciousness. These powerful emotional states are thought to have been experienced by tribal shamans while in Cascadia Cave. The long zigzag line inscribed in the cave and the crack in the wall indicate where the wall "opened up," according to an Indigenous religious leader. Shamans are believed to travel to the spirit world, where they are able to become an animal, and where an animal might transform into a person.

Back when the cave was actively used, about eight thousand years ago, the rock shelter was set above a clear meadow, theatrically set as the headstone of a giant amphitheater. Farque imagines a shaman leading a ceremony at the cave mouth and groups of people below in the meadow, drumming, dancing, and singing around ritual fires. This site was and continues to be significant to the local tribes as the link from the physical world to the metaphysical world. Ceremonies would emphasize healing and human responsibilities to renew the natural world. The Kalapuya people, native to this region, would come here to gather spirit power.

Native Americans may see animals as "persons," and even though they are hunted, the animals are still respected. Northern Alaskan Nunamiut tribes practice rituals to allow the "breath" spirit of their prey to escape, so it can tell other animals that the people had done well in their killing, and to allow the spirit to be reincarnated.[2] Animals are respected because they have their own forms of intelligence. Rituals around hunting and killing an animal helped keep the spirit of the animal alive. Native Americans believe in a spirit world that links people with the world of nature, with their ancestors, and with the future. Native American traditions have linked them with animals for ten thousand years or more in an unbroken continuum.

These insights help us understand the significance of cave paintings and artifacts from some of the earliest modern human beings in Europe. Many European cave paintings contain similar signs, like zigzag lines, to those in Cascadia Cave. Amazingly, they have been protected for tens of thousands of years in deep caves, and are older—some are four times older—than those in Cascadia Cave.

CRO-MAGNON ARRIVED in Europe and with other groups of *Homo sapiens* spread eastward, across Eurasia. These included the ancestors of Native Americans who traveled to North America. Cro-Magnon people lived in small, mobile bands. They had dark skin, were healthy, as tall as Westerners are now, and lived relatively long lives, according to their DNA and other information left in their bones.[3] Over the entire span of Ice Age cave paintings, these people painted similar animals, in similar styles, showing a consistent system of belief over time.

We visited several different caves in France based on geography, not on chronology, traveling from the western edge of the Pyrenees northeastward to the Dordogne valley. We used what we learned from Cascadia Cave to try to understand the significance of European cave art.

NIAUX CAVE

Our bus lumbers up the road in the foothills of the French Pyrenees until we reach Niaux Cave. We disembark and gather eagerly around the narrow, dark mouth of the cave, where we meet our guide, Françoise. Each of us holds a small, dim lantern that makes only a little impression on the otherwise total darkness. After about twenty minutes of fast walking in the dark, we round a bend where the cave opens into a high-ceilinged cave room called the Salon Noir. We are asked to turn off our lamps. The darkness is sudden and absolute. Françoise turns on one single beam. We follow her, taking a few more steps into the large chamber—and stop—facing a wall. Two meters in front of me is the bison with the luminous eyes.

The bison looks out of the rock wall at me with her shining eyes; so vital and beautiful, almost laughing. It has been fourteen thousand years since she was painted. I stand there mesmerized. The artistry is clever—the front leg of the bison shows greater detail than the smaller back leg, indicating perspective. Other overlapping figures show the artist's understanding of depth and proportionality. They knew how to draw images as they appeared to the eyes long before the artists of the Renaissance reintroduced it. The bison's outline was etched into the rock, making the image pop out. There are no sketchy lines; this and the other paintings were made with confidence.

We modern people do not stand at some pinnacle of human capability or intelligence. Tens of thousands of years ago, highly capable Cro-Magnon people created these skilled and evocative pictures of animals. In even the earliest paintings, animals are depicted with respect, even reverence, skillfully, in a way that indicates an understanding of the shape and behavior of animals. I question what modern human "progress" means and wonder what we have lost along the way. What important wisdom about avoiding ecological collapse can we still retrieve?

Ice Age peoples weren't merely surviving; they brought about a flowering of human culture. The Cro-Magnon took the time to reflect on life's magnificence, having vibrant relationships with nature. These evocative paintings are hopeful—there are none indicating war, avarice, or greed. They seem to have lived as equals; there are no signs of a hero or a king. The paintings of animals show such detail, are so tenderly etched; they reflect elements of their moral character—respect, reverence, reciprocity.

The Ice Age tundra has been called the Serengeti of Europe. Herds of large animals lived on the broad plains, including woolly mammoth, mastodon, and bison. During the summer, small bands of one or two dozen people lived in mammoth-bone huts covered with hides. The people probably followed the migrating reindeer herds. They ate reindeer and sometimes mammoth, red deer, or horse, and they caught fish.

In Niaux Cave, paintings were made and rituals were held; the people played music and probably even danced, to the light of flickering stone fat lamps illuminating the animals on the walls. The music and the paintings might have linked together "the visible and invisible worlds," suggests paleoanthropologist Ian Tattersall.[4] The spacious Salon Noir is shaped like a cathedral—acoustically ideal for transmitting sound. Françoise invites us to call out or sing while standing there under the high dome. Her soprano "Hmmmm" rings out and resonates for several seconds amid the animal drawings around us. These reverberations indicate how music was likely used here. There are many other caves with both acoustic properties and abundant animal paintings like this one. Small bone flutes have been found in cave sites from this time period, like Isturitz, made of vulture wing-bones, thirty-three thousand years old. Stalactites hanging from the ceiling appear to have been struck, and ring out when struck. Each stalactite has a different ringing sound; archaeologists call them lithophones.

Rituals probably fostered an unconditional social trust and group rapport. Anthropologist E. Richard Sorenson identified this level of awareness, based on experiential awareness, in his studies across different present-day hunter-gatherer cultures.[5] The people he studied somehow managed to isolate themselves from aggressive colonial influences. During rituals of transition, people experience a disorientation as they transition from one way of being to another. Rituals continue to provide humans with a sense of belonging and a connection with the past and future. We might experience a similar state of ambiguity ourselves while participating in weddings, or rituals around the death of loved ones.

Music itself helps bind people. Disa Sauter and colleagues suggested that humans compensate for language barriers when encountering different peoples by using music; certain rhythms, pitches, stress, and intonation are common across cultures.[6] Music can transcend language and help navigate situations of uncertainty between groups and with nonhuman forces. The paintings in these caves and the probability of musical rituals suggest a rich mental world. If they used music, we speculate that these people also had a complex language.

Millions of grazing animals lived on the tundra and steppes during the ice ages; the coldest periods were also times of greatest numbers of animals. Fossils provide evidence of the abundance of grazing animals that lived on the richly vegetated plains. The artists clearly paid close attention to these animals, observing every detail of light, color, shape, and motion. No bison lived there, although there is a concentration of bison drawings in the cave. This implies the bison lived in the artist's memory, carried to this site from elsewhere. These images convey a deep understanding of the animals. Every important detail was attended to—the angle of horn, the flatness or erectness of mane, how the animal moved—and reflected in the art. A painting of a horse shows whether it was male or female, gravid or not, living in spring or winter. However, these paintings were stylized, unlike the photographic images we modern people are accustomed to. The bison horns each look like a double-elongated *S*. The Salon Noir was painted over an extended period, and newcomers, several thousand years later, made their paintings in the same style as the first painters used. Stylization in the shape of the horns is one example of cultural transmission of technique. They evidently respected the past. Did they use the older paintings in the cave with different rituals than for the newer ones?

MODERN-LOOKING HUMANS ventured into Europe about forty-five thousand years ago, traveling up from Africa through the Middle East. These people, the Cro-Magnon, lived during the Paleolithic era, a time when temperatures varied between 12°F colder than present during glacial advances and at least 2°F warmer than present when glaciers retreated. Pollen evidence tells us poplar and ash trees grew in some places, but most of the land was an open steppe-tundra. The glaciers advanced and retreated as many as thirty times; these were broken into four major periods of glacial advances (maxima) and retreats (minima). Four distinct but related groups of Cro-Magnon moved into Europe over this time, corresponding to the major glacial minima and maxima.

The first culture, the Aurignacian, began when Cro-Magnon people first moved into Europe about 40,000 years ago. Their art included a small statuette of a horse, a lion-human figure, and the paintings of Chauvet Cave. When glaciers advanced again, these people moved south into what is now southern Spain. Different but related people migrated in from the east during the Gravettian, 33,000–22,000 years ago. They left distinct artifacts, cave art, and musical instruments, including the so-called Venus statuettes, an elegant ivory carving called the Lady of Brassempouy, the handprints of Gargas Cave, and the painted dappled horse and handprints of Pech-Merle Cave.

Ice sheets then expanded across northern Europe and down from the Alps and other ranges during one last glacial maximum. Humans contracted southward into small favorable habitats or refuges. This period, the Solutrean, lasted from 22,000 to 17,000 years ago. This was when people created the renowned Lascaux Cave, and distinctive "laurel leaf" spearheads were knapped. As the ice began receding for the last time, people similar to the first cultural group, the Aurignacian, moved up from what is now Spain into what is now France and Germany. This period, the Magdalenian, lasted from 17,000 to 12,000 years ago, when Ice Age art was at its peak. The people decorated many different materials—bone, stone, and ivory—each in a different manner. Their cave art included Niaux, the drinking reindeer of Combarelles, and the renowned black and red bison of Altamira. Although it was still very cold, by the end of this period, a long-term warming trend had begun.

The act of making an image deep underground involved crawling through narrow, sometimes very long, dark passages to the underworld. Group rituals were performed in some painted caves, as if the art was made to be performed. Mysteriously, most painted caves were rarely visited, evidenced by the few footprints left behind; and when they were visited, only a few people came. The most remote paintings were not seen by anyone other than the artist. In Niaux there is one very remote cave where one artist entered, left footprints, painted a bear, and then turned and left. No other footprints are present. The private act of making paintings must have been a ritual connecting the artist and the animal symbol. Perhaps these remote underground paintings were thought to help make the mountains and the landscape sacred. Perhaps they were part of shamanic rituals. The cave walls may have been perceived as membranes between the known and the unseen worlds; the act of touching the walls allowed access to the world beyond.[7] As I exit the cave, the fresh air smells sharp in contrast.

WE TRAVEL ACROSS the valley from Niaux to visit the Grotte de la Vache, where the cave painters lived. Engravings on bone, ivory, antlers, or flat pieces of stone were found here. One scapula was carved with a running lion engraved on both sides, with the eye in the same position, but the body slightly moved to represent phases of a running lion. The guide hands me a facsimile of the engraved bone; when I flip it, it appears to move. It seems to have been made to tell a story in the manner of a modern flip book. An engraving made on a whale bone was found nearby, forty kilometers from the sea. A pendant, carved with a figure representing a sperm whale, was also found. Perhaps some of the group had moved to the milder coastline in winter. Other types of animals were carved on femurs, scapulae, other bone surfaces, and ivory was carved into beads, fox teeth into necklaces.

To the north of the steppe-tundra grassland with herds of grazing animals were the large continental glaciers covering northern Europe. Just south of the glaciers themselves stretched a periglacial desert, where dry winds blew, sweeping up a fine dust ground by glaciers, called loess, over long distances. In places, loess accumulated tens of meters thick. The loess dust eventually buried and preserved both human camps and animal bones and carcasses of the food they ate. The preserved remains provide insights into how these people lived.

CAVERNE DU PONT D'ARC

We travel north to visit Chauvet Cave, although we are visiting a replica because we cannot enter the actual cave. Archaeologists worry that human breath, which caused serious deterioration in Lascaux Cave, might destroy the precious art here too.

When Jean-Marie Chauvet and his friends discovered this famous cave in 1994, one of them exclaimed, "They were here!" Another described feeling how "time was abolished," explaining how the tens of thousands of years that had passed since the artists created these paintings seemed to evaporate.[8]

Inside the replica, the texture of the walls, the stalactites, even the interior temperature and humidity of the cave are meticulously imitated. Knowing this is a replica, I am at first unimpressed. We come to the panel of horses; four magnificent horse heads are drawn, each superimposed on another. The top one, with an upright mane, has its muzzle pointing slightly upward; the second has its muzzle and head pointing down; the third slightly upward, and the fourth more upward still. Taken together, they look like an animation of one horse tossing his head. Other art in the same cave has overlapping engravings of the same animal; the panel of lions shows each of three lion heads somewhat superimposed over another, seeming to depict movement.

These powerful and extraordinary depictions of animals were drawn very early on in the era of Ice Age art, between 33,000 and 36,000 years ago. The dates were confirmed using accelerator mass spectrometry radiocarbon analysis of charcoal used in some of the drawings. In this earliest of the painted caves, the artists used all of the artistic conventions and depicted most of the animals subsequently painted throughout the Ice Age.

In this and in many other painted caves, dozens of different geometric signs were inscribed. There are twice as many geometric signs as there are paintings, says Genevieve von Petzinger, who tallied all of the signs in caves.[9] Straight lines, feather-shaped forms, club-shaped signs, arrow-shaped lines, curvy and zigzag lines, series of dots, triangles, spirals, crosshatching, asterisks, and ladder shapes appear in the various caves. Several signs are often found together, such as dots with negative handprints. Patterns of dots painted near passageways could have been passage markers, such as "turn left here." Some patterns of dots coincide with places where a tunnel has maximum resonance, as if marking the way to a sounding spot. A few of the sets of dots could have represented patterns of star constellations—important to hunter-gatherers because the appearance of constellations in the sky indicates times of year when herds migrated. On the ceiling in Gargas Cave, waving fluting patterns dug in the rock suggest the real-world pattern at the nearby confluence of the Neste and Garonne Rivers near the cave.[10] Symbols placed there could have depicted other aspects of the landscape, with triangles standing in for mountains. Many signs in the caves appear in some relationship with the paintings.

If the paintings were made in connection with the spirit world, the act of painting, etching, and rendering a sign over the animal might have been an act of magical interaction with animals. French archaeologists Julien d'Huy and Jean-Loïc Le Quellec speculate these people were interacting with the procreative forces of nature from which the animals and maybe the people themselves arose.[11] Some of the bison painted in Niaux Cave were overlain with upward-facing arrow shapes. Even though early archaeologists believed these signs were depictions of spears that would magically ensure hunting success, modern evidence provides new interpretations. The paintings are more magical than that. The artists rarely painted reindeer, which their people hunted and ate. Bison were painted at Niaux even though there were no bison there at that time. There are dark lines enigmatically drawn into the animals. These lines might have been a way of asking the spirits to "come forth!" Reenacting a myth of how the animals may have first emerged from the walls of the cave, the people might have been ritually trying to replenish their world with animals. Archaeologists say this would explain why the dangerous animals were always drawn deep in the recesses of the cave; for example, a woolly rhino was drawn in a deep shaft. According to the myth of emergence, dangerous animals might have been painted so deeply back in the cave to spatially delay their emergence until the artist moved out to safety.

Some pieces of portable art, made on bone or pieces of stone, have engraved marks that could represent ancient timekeeping. Alexander Marshack examined sets of tiny marks made on bone tools.[12] One of these formed a twisting pattern. He reasoned that these marks were a type of notation for timekeeping. He found that series of marks on some bones had been made over time. One bone with seventy different, separate marks may have been created as a moon calendar, possibly depicting the changing moon over two consecutive months. I wonder, why was timekeeping important for the maker of the calendar?

GARGAS CAVE

At the entrance of Gargas Cave, in the Neste river valley, we enter via a steel door set into the limestone wall and walk some distance through the upper cave tunnels to the lower cave. There, on the cave wall, is an array of 231 black handprints, made by a group of twenty to forty individuals. A wave of excitement passes through me. These prints are made from direct contact, from someone's actual touch from long ago. A child's handprint is high up on the wall; someone—possibly a father or mother—must have held the child up and guided its hand to make the print. There is tenderness in this gesture. Most are "negative" handprints, made by holding a hand against the wet surface of the cave and spitting or blowing powdered pigment around the hand through a bone tube. Some were made using a mossy pad with pigment sponged around the hand. Most were made with charcoal but a few with red ochre. The pigments have been found in archaeological layers dated to between 27,000 and 24,000 years ago.

Similar handprint images and petroglyphs have been found in other places around the world: Argentina, Borneo, Australia; and in the United States, in Nevada, Arizona, Utah, and Oregon. The handprints must have been symbolic—a means of identifying one's self to others, or a personal link with the forces within the rock walls?

Perhaps the handprints were a coded language. In Gargas Cave, many prints indicate the fingers of the hands had been folded in, as if the hand were missing a finger or two. By bending the fingers to create a coded meaning, the artists used a sign language whose meaning we cannot now know. In this cave, handprints are concentrated in one part of the cave in a panel. There is a single baby's handprint near a painted image of a vulva, as if indicating the source of new life. In some cases, the handprints are placed along passageways, as if serving as beacons to guide people through the cave.[13]

ICE AGE PAINTINGS arose out of a belief system in a richly mythic world. Their belief system must have supported a need to create and use paintings in some ritualistic way. Even though each culture is unique, with a range of behaviors, we can look to the belief systems of modern hunter-gatherers, which can show us possibilities and suggestions of what might have been then.

Modern hunter-gatherers have different ways of perceiving themselves than do people living with modern Western conventions. Some understand themselves as permeable and continuous in the world of nature. For example, many Aboriginal Australians believe that when they die, their spirit returns to the earth, where they join long-departed ancestors. Ancestors can be accessed by speaking directly to them, says historian Richard Broome.[14]

Ice Age people might have believed the spirits of ancestors and animals could influence the events of living humans. Ice Age hunters may have had an affinity for particular animals and their spirits. Ice Age people's cave paintings and handprints could have been a means of communicating with the animals' or ancestors' spirit world through shamanism. These beliefs and other aspects of human consciousness moved around the world along with *Homo sapiens* as people migrated around the globe.

The handprints like the ones in Gargas Cave could have been a way to touch the veil between the worlds, speculates anthropologist David Lewis-Williams.[15] The cave walls themselves were thus an interface between distinct realms. French prehistorian Jean Clottes believes prehistoric shamanism is the best explanation for Ice Age art.[16] A shaman is believed to send their soul out of their body and communicate with the forces that direct human life, represented by powerful animals. Shamans met with these spirits because the people needed their help. For Upper Paleolithic people, crawling into an underground cave may have been an act of moving between worlds. The art arose from something beyond the level of ordinary existence; most likely the images were influenced by shamanistic practices. The vitality of the animal paintings and carvings communicates a sense of the transcendent to many viewers and archaeologists.

French anthropologist Pascal Raux thinks some of the decorated Paleolithic bones carved with animal forms were involved in a shaman's ritual journey to the spirit world, similar to the crack in the wall and the zigzag line in Cascadia Cave.[17] Several of these bones have a large carved hole. One depicts a buffalo head emerging out of the hole. The image appears to echo a drawing on the wall at Niaux. Dr. Raux thinks this carving shows the power of the shaman to make the transformation to the spirit world and to return again.

Although human figures on cave walls are rare, there are four well-known engraved figurines and paintings from the Ice Age of human figures that appear to some archaeologists to depict shamans. There's the bison-man engraving and the painted "sorcerer" in the Trois Frères Cave. The sorcerer figure is a strange combination of a deer body with a human face. It was named the sorcerer by Henri Breuil, an early archaeologist who visited the cave a century ago. There is a painted bird-man figure at Lascaux in a hidden shaft of the cave. It depicts a sketchy human figure with a birdlike head and an erect penis facing an enraged-looking bison with its entrails pouring out below. There is also an ivory lion-headed man sculpture from Germany.

Human figures engraved on bones and portraits of men and women's heads engraved on stone plaques have been recovered from a few caves. Several thousand plaques have been found in one cave that were apparently brought from elsewhere. They are odd and very different from cave paintings.[18] Many depict lifelike human figures with overlying squiggly lines of reddish pigment. They have been dated radiometrically from the Magdalenian. Most of the male faces are clean shaven; some have mustaches or trimmed beards. One stone depicts an expressive-looking woman with a pointy nose and wearing a cap. The stones show images of humans wearing tailored clothes or heeled knee-high boots. Some stones contain characters who are dancing, and one shows a woman and a newborn infant. We might prefer the realistic images, but French archaeologist Oscar Fuentas believes the schematic or distorted images are more "advanced" because they transition from the literal into the symbolic.[19] Inexplicably, there are squiggly lines superimposed over most of the sketches. No one today knows what those overlying lines mean.

The stone plaques show us what Cro-Magnon people wore. They sewed and wore tailored animal-skin clothing with patterns of beads and also adorned their bodies. Besides the images of clothed people sketched onto stone plaques, there are shells patterned with carvings found on buried skeletons thirty-seven thousand years old and bone needles found in the cave debris. And there is a tiny, delicately carved ivory ornament, called the Lady of Brassempouy, which depicts the head and neck of a woman wearing a net over her long hair.

Archaeologists interpret the time when cave art was created as a period of rich surplus. Cro-Magnon had a rich culture, an abundance of food, yet few material possessions. Their lifestyle had a sort of Zen strategy, according to American anthropologist Marshall Sahlins. He interpreted their lives as having rich affluence without the baggage of possessions.[20]

CLIMATE CHANGE ultimately led to the collapse of Ice Age cultures. As the Cro-Magnon's world warmed, forests encroached into the once widespread plains, and large game animals disappeared. A warmer geological period began—the Holocene—marked by an unusual period of stable, dependable climate and flourishing wild cereal grasses.

At first, people gathered cereal grain to complement food caught by hunting and by foraging wild foods in Europe. People settled down in ecologically rich areas and began sedentary living. Gradually, over a period of about five thousand years, a farming way of life spread along the Mediterranean coast and up the Danube valley, eventually becoming established in Europe.[21] As farming and animal domestication became established, so did infectious diseases, and there was a general decline in health. The bones from this later period, the Neolithic, show malnutrition and a shortening of people's stature. Remains from farming villages contain evidence of inequalities between groups of people and the rise of privilege.[22] Paleogenomic researcher David Reich calculated the invasion of nomadic herders from the steppes of eastern Ukraine and southern Russia to about five thousand years ago. These people brought the practices of warfare along with them.[23]

Today, we live in a new geological era marked by decreasing stability in climate and increasing uncertainty about the future. Cro-Magnon disappeared, yet their Paleolithic DNA is part of the makeup of people with European ancestry.[24] Many Europeans continue to feel linked to the Ice Age people. "They are our ancestors," one cave guide told me. French schoolchildren are taken on class trips to witness the cave art as part of their patrimony.

ICE AGE ART IN OTHER CAVES

Ice Age art is not confined to Europe—it is a widespread phenomenon. Some of the same symbols are used in many other places around the world. The combination of paintings and symbols could possibly represent early writing.[25] The tectiform—an upward-pointing arrow—is found at fifteen different sites in European cave art and in Africa and Australia. The odd sideways-lying ladder form called the scalariform is found at three sites in Europe, such as Altamira in Spain and in the Dordogne in France, and in Blombos Cave in South Africa. The spiral and negative handprint are also found in many other places. Negative handprints are found in caves in twenty-three different sites in France and Spain, and were made from the late Paleolithic era up until the early Holocene or modern era. Negative handprints are found around the world, but are especially known from Gargas Cave in the French Pyrenees, in Ubirr in Australia, La Cueva de las Manos in Patagonia, Argentina, China, North Africa, and in North America.[26]

The earliest designs ever found—carved onto pieces of bone and rock colored with red ochre—come from Blombos Cave in South Africa and are between 70,000 and 100,000 years old. Engraved seashells, probably worn as ornamental beads, have been found in South Africa, North Africa, and the Near East; some are 100,000 years old. A perforated tiger shark tooth was found in New Guinea and dated to be 38,000 years old. Painted and engraved animal images 28,000 years old were found in Namibia. Anatomically modern human beings in different parts of the world began to engrave and paint symbols and images a long time ago, but especially at the end of the Pleistocene.[27]

AFTER VISITING THE PYRENEES, Roger and I travel to Font-de-Gaume Cave in the Dordogne valley. There we see the drawing of swimming stags, represented so that their heads are arched upward out of the water. They look as if they are swimming across a deep river; the landscape is conveyed by a brown layer of rock just below the stags' shoulders and with splashes on the wall looking like water drops.

Later that day, we visit Les Combarelles Cave, near the famous Lascaux Cave. The Ice Age artists would have had to crawl down a narrow three-foot-high passageway, using only the light from fat-burning lamps to see with, to paint the images, our guide Stephanie tells us. The floor of this cave was recently excavated out, so we are able to walk easily down the passage to see the carved animals on the walls.

Inside the cave, Stephanie positions her light from below, as the Ice Age artist might have done. This accentuates the lines etched into the limestone cave wall. In the light halo thus formed, we see *La reine qui boire,* the engraved drinking reindeer. Perfectly proportioned, its head bends down and out toward a natural seep in the cave. Graceful and whimsical, it is etched so that its tongue is protruding as if it were drinking from the seep in the rock. Light and shadow—created by rocky protrusions and concavities—emphasize the form of the animal. As before, in Niaux Cave, my gaze fixes on the reindeer's animated eyes. Gazing into the expressive eyes of one of these animals, something ineffable is transmitted. Imagining those eyes, I feel a sense of sorrow and compassion for our present human plight.

As THE ICE AGES were ending, some groups of people moved out of Asia to the Americas, moving along a glacier-free coastal area. This ice-free zone was also a refuge for plants and animals. Many archaeologists now believe a sea voyage helps explain the early settlement of the Americas. It is thought people immigrated into the Americas 18,000 years ago and lived in isolation in the north until 15,000 years ago, when the Pacific coastal corridor became deglaciated. The first Americans sailed down this newly opened-up route.[28] At the time, the sea level was hundreds of feet lower, and many estuaries rich in seafood would have been available to the newcomers. This coastal migration theory is based on genetic, archaeological, and environmental evidence.[29] The interior corridor, previously thought to be responsible for the first migrants' passage to the Americas, was probably not ice free until a few thousand years later and cannot be used to help explain an earlier timeline indicated by new findings. Archaeological evidence that humans were present as long ago as 14,600 years has been found at the Hebior site in Wisconsin, at Monte Verde in Chile, and in the Schaefer site near Chicago. Humans were also living at Meadowcroft in Pennsylvania, Page-Ladson in Florida, and Paisley Caves in Oregon around this time.[30]

Sets of footprints, possibly as old as 14,000 years, were recently found pressed into a layer of brown clay on the shoreline of an island in British Columbia. Small bits of wood found pressed into the heel of one print allowed them to be radiocarbon dated.[31] On reaching the Pacific Northwest, people would have continued their spread southward along the coast to Chile, and some might have moved eastward, out below the southern margin of the continental ice sheets. They might have followed traces of mammoth and mastodon as far east as present-day Wisconsin. Petroglyphs depicting Pleistocene fauna and bighorn sheep found in the Mohave Desert, aged by the rock varnishes that now cover them, could have been carved into the rock before 12,000 years ago. Older sites in the Pacific Northwest might exist, but have been inundated by rising sea levels and buried by deposits of sediments from the series of Missoula floods. Earthquakes and tsunamis probably destroyed and flooded older and lower-lying villages.

"Our modern skulls house a stone age mind," Leda Cosmides and John Tooby remind us.[32] Paleolithic priorities have shaped our brain to be far better at solving some problems than others. It's difficult for us to perceive slow change over time. Yet our Paleolithic brains give us hope. We are hunter-gathers wearing suits and ties, dresses, jeans, and T-shirts, says Simon Barnes.[33] Ice Age people seemed to live in awe of the life surrounding them. They seemed to know where they stood in the scheme of things. The stories we tell each other about where we come from make a difference in how we live our lives. Considering our links to Ice Age ancestors and using examples of modern Indigenous peoples' reverence for the natural world, we can become better advocates for healthy ecosystems. Remembering how the world was made not only for us, more of us learn to accommodate the wild.

CHAPTER 3

Standing Stones

THE PLEISTOCENE ENDED. The ice receded as temperatures warmed. The megafauna, mammoths, mastodons, woolly rhinoceros, cave lions, and other large mammals disappeared. Grasslands spread out. Hunters and gatherers began collecting wild grass seeds to supplement their diets. People in ancient Mesopotamia settled down near fertile estuaries and experimented with flood-retreat agriculture.[1] This involved scattering seeds on the rich ground recently vacated by receding floodwater. When the water rose again, it covered and irrigated the seeds. The grains harvested were a minor part of these people's nourishment; gazelles were hunted and marine food and freshwater fish were eaten. In some places, like the island of Cyprus, when hunting and farming people arrived, they deliberately brought with them free-living deer and foxes as well as domesticated herd animals. They hunted wild animals, farmed, and raised livestock.

Small-scale farming experiments like those continued for thousands of years. As the human population grew, the flat floodplain areas under cultivation weren't enough, and agriculture crept up the hillsides. Erosion began to wash topsoil down into the river deltas.

The agricultural "revolution" was not a glorious march forward for civilization as has been portrayed. In fact, it's been called "history's biggest fraud."[2] People worked increasingly longer in the fields to grow and harvest grain. Lives became harder as human cultures relied increasingly on grain. Farmers had shorter lives, calculated as a maximum of thirty-five years for men and thirty for women. Their decayed teeth and shorter stature indicate nutritional deprivation. Neolithic women bore more children than their Paleolithic predecessors. Childhood mortality increased. Women's bones show skeletal deformities associated with hours of daily grinding of grain. Once harvested, grain needed to be stored, and stored grain became a source of wealth. Some societies became more stratified as surpluses enabled the development of well-fed elite classes of people but poverty for the others.[3]

The archaeological records from those times show cyclical population booms and busts. After a boom, and fall, people returned to hunting and gathering. After some time passed, farming experiments picked up again. This cycle repeated over thousands of years. Journalist Richard Manning thinks early farmers preferred foraging and hunting because it afforded them more leisure time. Farming, in contrast, was backbreaking drudgery.[4] Manning suggests that each subsequent return to farming might have been coerced by a group of elites, who forced people back into farming.

As the scale of cultivation increased, another shift occurred. The older reverence for wild animals was supplanted by a reverence for human ancestors. Ancestor worship indicates a profound cultural change toward human-centeredness. Many cultures moved toward subjugation of animals, according to French anthropologist Catherine Perlès.[5] This shift was accompanied by a shift in human social arrangements, the rise of hierarchies. Many hereditary leaders acquired heightened prestige by linking themselves with ancestral spirits. Chiefdoms concentrated wealth and power. Societies continually collapsed and changed. In many places in the world, kings arose who claimed themselves to be uniquely empowered by a divinity.

We visit a site in France dating from the beginnings of farming in western Europe.

CARNAC

Carnac is a giant megalith structure erected on Brittany's coast over five thousand years ago. Constructed fifteen hundred years before Stonehenge, these arrays of thousands of standing stones also contained human bones buried beneath the megaliths. These farmers apparently worshiped their ancestors.

Roger and I arrive by train and stroll through the high-arched stone gates of Vannes to its old medieval center. Half-timbered two- and three-story buildings with wooden crosspieces, now painted a colorful red, orange, green, or brown, line the streets. Our room for the night is in a rustic two-story stone building just outside of town. The French doors of our rental open out onto a meadow bordered by a distant stone wall. We sit in wooden lawn chairs out in the meadow and gaze into the twilight. Beyond the stone wall spreads a deciduous forest extending far into the distance. The evening is green and peaceful.

FARMING PRACTICES, seeds, and animals were brought to France by new groups of European immigrants around seven thousand years ago. The DNA of the people who settled in Brittany contains a mix of genes from Turkish-Anatolian immigrants and local Indigenous hunter-gatherers.[6] Here, they engaged in small-scale farming while also hunting, gathering, and foraging in the marine zones for rich coastal resources over several thousand years. The farmers' bones bear the signatures of marine proteins from seabirds like auks, seaweed, seals, shellfish, and fish. The bones and shells from their meals are found in middens, along with bones from wild boar, hare, beaver, migratory birds, and fruit pits. Buried in these middens are burnt wheat grains and grinding stones used to process the grain. Primitive sickles for harvesting grain, with curved wooden shafts holding sharp toothlike flint implements inserted inside the curve, were also found.

WE SET OFF the next morning for Carnac. There we meet our guide, a tall, slender young man wearing dark-rimmed glasses and with a serious demeanor. Samuel works for the Musée de Préhistoire de Carnac. He explains our itinerary and gives us some stone vocabulary. *Menhir* means stone in Breton, *cromlech* means large standing stones, and *dolmen* means stone table. *Cairn* means a covered dolmen—usually a tomb that was covered with earth and stones. Megalith consists of two words—*mega* is Greek for large, and *lith* is Greek for stone.

Megalith building itself seems to have originated in Brittany. Using radiocarbon data from more than two thousand samples, researchers have traced the spread of megalith architecture from Brittany along the seaways of the Mediterranean and Atlantic coasts in three different pulses.[7] These megalith builders themselves might have been seafaring. Alternatively, it might have been the culture of constructing megaliths that moved and not the people themselves.

Tens of thousands of megaliths are scattered throughout Europe; some were erected sixty-five hundred years ago. Thousands occur in this area. The standing menhirs range in height from three to twelve feet. Carnac has a series of three different alignments, each organized in similar ways. Single stones also dot the landscape. Some mark monumental graves. Other megaliths stand in circles. Most of these rocks had been naturally eroded away from nearby bedrock outcrops, with only minor quarrying needed to extract them from their source.

Communities of an estimated forty thousand people lived near the alignments.

Carnac's remarkable megaliths were erected in long undulating lines of standing stones. They are aligned in parallel rows of solemn, weathered granite—four miles long. The gray granite stones, now covered in gray-green fuzzy lichens, stand along eleven rows. They were placed so as to increase in size from east (the direction of birth) to west (the direction of death) along the flat field. Some stones are irregularly shaped, with a big belly or with a small foot and large head; some are rounded, some sharply pointed. At the western end stand the tallest stones, about thirteen feet high. At the eastern end of the rows, where menhirs are shortest, around three feet tall, is a large ring of stones. The ring probably once enclosed a Neolithic temple. Now within the ring stands a group of relatively modern homes with slate roofs.

As we walk along the rows of stones gleaming in the morning light, the lines of stones seem like an outdoor cathedral. Neolithic people probably held rituals here during the solstices. The ritual might have been seen as assisting in the rebirth of the sun on the shortest days of the year. Samuel points to the tall cromlech standing at what was once the ritual entrance of temple: "We think Neolithic people used these alignments in ceremonies," he says. "People would walk in procession along them as they headed toward the entrance to the temple."

Inside the now long-gone temples, the living might have held rituals to ask their ancestors to intercede on behalf of the living. Praying to ancestors is still practiced today in different cultures throughout the world. The megaliths themselves, along with buried ancestors, might have been perceived as the guardians of the land, anchoring people to this land. Some megaliths might have even been painted, since painted and decorated items dating from those times have been found.

Here at Carnac, the acid soil destroyed any buried human bones, but bones have been found in nearby sites of the same age. In some places, the bones indicate the remains had been buried, revisited and unburied, then reburied, over several thousand years. During these revisits, people left new tools and other offerings to the tombs. Bones found at some English Neolithic tombs were bleached white, indicating they had been removed from the grave, bleached in the sun, and set back in the grave with other bones repeatedly over long periods of time. Archaeologists David Lewis-Williams and David Pearce speculate about how the bones of the dead were brought out in powerful ritual displays involving dance, music, and storytelling about heroic ancestors' deeds.[8] The rituals tied the landscape to a mythology binding the people to their ancestors.

At first, a group leader was probably seen as one among equals, since early burials do not contain particular prestige items. But farming and the requisite storage of grain after harvesting led to a lopsided accumulation of wealth. Burial chambers hold evidence of heightened prestige for the few interred there. Food storage vessels and precious ornaments such as axes made from jadeite from the Alps and turquoise-like variscite beads made from stone from Spain were found there.[9] This evidence shows how hereditary leaders benefited from an unequal distribution of wealth.

After walking around Carnac all morning, we are hungry and head to a crêperie in town. Once seated, I order a crisp buckwheat crêpe filled with cheese and tomatoes. It's delicious. Roger and Samuel drink mugs of cider along with their crêpes. Crêpes are important to Brittany. This region was independent until absorbed by the Kingdom of France in 1532. After the French Revolution, Breton people were pushed to the poorest agricultural lands. They could grow only buckwheat and a few other crops. Ironically, buckwheat crêpes are now fashionable and are served in fancy crêperies in Paris. Buckwheat from Brittany carries its own protected French geographic designation (IGP), emphasizing how it has been cultivated and milled using techniques specific to this region since the fourteenth century. This designation highlights how buckwheat is still farmed here without using chemical weed killers.

Samuel tells us about stone burial chambers and earthen-covered burial mounds called passage tombs built by the people who constructed Carnac. After lunch, we visit one of the passage tombs named Kercado. Constructed in 4600 BCE, it is a tall, long mound—forty meters long and twenty meters tall at its highest—with a dolmen erected at the top. It was covered with rocks and earth and encircled by a ring of stones. We duck down to enter the passage and walk in the semi-darkness for about six and a half meters, entering a large chamber where we can stand upright in the center. Engravings on the large stone capstone covering the chamber—called "axe-plow"—consist of long horizontal rod upcurved at the end with a downward facing triangular piece hanging from it. Inside the tomb it is solemn and reverential.

Samuel explains how passage tombs like this one were constructed to give the impression of entering another realm. Once you enter this realm of the dead, you stand in the center of a circular area representing a re-creation of the cosmos. Inside the tomb, it is silent and dark as it would be in the underworld of the dead. These earthen and stone tumuli were fashioned so that the entrance and the tunnel leading in or out faced the setting sun during the solstice. During the winter solstice, the rays of the rising sun would penetrate to the back of the tomb, illuminating the large stone at the back of the burial chamber. Chris Scarre and colleagues believe the path of light, leading from the underground burial out to the sky, would have dramatized for living witnesses the spiritual journey taken by the dead.[10]

The people believed themselves tied to their ancestors who were buried in a stone tomb. The ancestors could travel from the underworld upward to the sky each year. These beliefs arose from their worldview. The dead were buried in the same ground that grew their crops, the earth the center of worship, the people tied to the land. The buried ancestors might have bolstered claims as to who owned and inherited the land and its wealth, say anthropologists Lewis-Williams and Pearce.[11] Believing in the power of these ancestors might have helped keep their cosmos in order.

This tumulus was continually visited over thousands of years. People might have come here continually to ask favors from the dead, even asking for favors from their dead children, over this long period of time. They might have venerated the dead out of fear of their powerful otherworldly status. At this site, the acidic soil destroyed any bones left, so there might have been multiple burials. Eventually, this tomb was sealed off. Other tombs, with single burials, were filled in right away—with no one apparently revisiting them.

AT LOCMARIAQUER, about thirty minutes away, we see three huge chunks of what was once one enormous granite block over twenty meters long. This is the Grand Menhir. After about two hundred years, it toppled and broke, either due to its own mass or an earthquake.

Now it sits on its side, broken into three pieces, weighing 350 tons. A fourth piece was moved to the nearby tomb of Locmariaquer.

The entrance to the tomb faces east, to the rising sun. Inside is the missing fourth piece of the Grand Menhir, placed as the capstone. I marvel at its axe-plow engravings and engraved animal shapes with four legs and long backward-curving horns. At the back of the tomb is another large stone engraved with many candy-cane crook shapes arranged in a pattern around the edges. This back stone was also once an outdoor self-standing menhir, evidently pulled down around 3,800 years ago and reincorporated into this passage grave. Archaeologists think these menhirs were moved by a new group of immigrants who had arrived from the south. Restructuring these megalithic monuments could have been a way the newcomers mixed with or imposed their culture on the existing one.

The engraved symbols on the stones have mythological significance. The axe-plow shape on the capstone has been found only at coastal sites. Some archaeologists interpret it to be an abstract engraving of a sperm whale. It appears on megaliths throughout Brittany, indicating a mythic connection with the sea and a continuance of ancient beliefs.[12] A whale image might have told the story of how whales transformed into the land or into the people themselves. Alternatively, whales might have been "clan" animals, representing the kinship prestige, stories, and prerogatives of a family. Whales continue to be powerful symbols today. Far away in the Pacific Northwest, Tlingit tribes still carve whales into totem poles to remind the people of their ancestry. Indigenous peoples continue to recognize how they receive gifts from these creatures, and reciprocate. Tlingit stories involve people transforming into whales and whales transforming into humans. I imagine how the ancient Bretons once did the same.

The candy-cane crook shape, on the stone at the back, was once thought to be a shepherd's symbol. But now it is thought to represent a throwing stick. Similar to a boomerang, this weapon was used in warfare and to hunt birds. It might be part of a mythic story about the action of a king, or a myth about the sky. The four-legged animal engraved on the capstone represents cattle and maybe represented earth itself. Evidence of charcoal found at the base of the stones shows that fire was also used here, most likely set during rituals. The different symbols taken together seem to reflect myths about creation: water, air, earth, and fire. The origin stories important to the farming new-comers and those of the Indigenous hunter-gatherers might have been combined.

These symbols, in particular those engraved on the capstone of a tomb, might have been intended for the dead. Lewis-Williams and Pearce suggest the journey of the dead was part of the cosmology of Neolithic people living in western Europe and Britain. The journey starts when the deceased is taken through the passage to the underground tomb. Archaeologists speculate that the living believed the dead traveled up through the capstone of the tomb, through the mound's roof and up into the sky. There, they became part of celestial events, like the solstices. The people might have believed their ancestors were accessible during the solstices, much like at today's Day of the Dead celebration in Mexico, held near the fall equinox.

Stone monuments like these continued to be elaborated and trans-formed throughout the Neolithic in western and northern Europe. None of these stones were deeply embedded in the earth. Modern-day schoolchildren have participated in reenactments of erecting big stones, with adult help. Ten people working together can raise a stone using wooden logs, a trench, and a hump by pushing the stone up to an erect position, then filling in the trench.

Lone megaliths stand on prominent coastal spots in Brittany, per-sistently oriented toward the rising sun. Along the Brittany coast, every-thing shifts with the tides—nothing seems permanent. Archaeologist Chris Scarre believes the large standing stones on coastal promonto-ries had a particular mythological meaning for the people.[13] This area has extremely large tidal ranges; large difference between high and low tides affect how you perceive the landscape. When the tide is out, large expanses of mudflats appear. When the tide is in, the mudflats are submerged by the sea. Twice each day, the tides cause these dra-matic changes, creating a sense of ambiguity between the known and unknown. This daily transformation—liminality—makes for a daily threshold between the visible world and the invisible one populated by spirits, says Scarre. Many tombs are located near the ocean where, with its twice-daily extreme tidal changes, a different and daily tran-sition between life and death would have been dramatized.

SKARA BRAE

About a thousand miles to the north of Carnac sits the ancient village site of Skara Brae, on an island off the northern coast of Scotland. We travel to the Orkney Islands to observe the site and imagine the ordinary lives of Neolithic farmers five thousand years ago. Neolithic farmers there built a community of houses out of stone, each with a stone bedstead, stone shelves, and stone hearth. These people also erected ritual stone circles nearby.

The ancient village lies inside a bay on the west side of the island. At the time it was inhabited, the ocean was far away, but now the ocean lies nearby. Today, a blue sky contrasts with the emerald green grasses surrounding the remains of the village. The thrill of seeing this place, which I had heard about many years before, passes through me. It had been protected by a deep covering of turf over thousands of years. Then, a succession of storms about a hundred years ago stripped away grass and sand to reveal the remains of the stone buildings. The village had been inhabited for six hundred years until it was mysteriously abandoned.

Standing at a viewing platform, I look down into the dry stone walls that enclose a cozy set of homes. They are ordered and well engineered, sunken into mounds of earth. Eight simple dwellings are visible, each one with stone beds, once filled with straw or heather mattresses. In the center of each home is a large hearth; along the walls are flagstone shelves, a sort of dresser, shelves for cupboards, and seats. Some storage boxes were found, sealed watertight. These boxes may have served as custom-built tanks for fresh fish and seafood. Networks of drainage channels line the village; they may have been engineered for primitive flush toilets. Paint pots with red ochre were found, probably used for ceremonies involving birth and death. Ivory needles, stone and ivory balls, and assorted colored beads made from sheep bones, cows' teeth, orca teeth, and boars' tusks were among the other items found here.

The homes are linked by a series of low, covered passages. Each home is equal in size, indicating a degree of social equality. Although they are no longer roofed, they most likely had pitched roofs lined with seaweed, animal skins, or thick grasses. The roofs might have been supported by timbers or whale ribs. The midden piles surrounding the houses show that the inhabitants were marine foragers, harvesting fish and shellfish. The people also grew barley and wheat, fertilizing their fields with manure. They raised sheep and cattle, but deer bones found in the middens indicate they also hunted.

A woman's skeleton was unearthed here; she had been healthy and stood taller than five and a half feet. The woman probably lived here, raised a family, and maybe even stood near where I am standing now, gazing out toward the sea. It feels as if very little time has passed since they lived here. The feeling is fleetingly hopeful. Perhaps we can survive the times in which we now live.

The Orkney Islands might have been the center of the Neolithic north—home to as many as ten thousand people. Pottery remains are some of the oldest of a type—grooved ware—found in Britain. Archaeologists Colin Richards and Nick Card believe the style might have started here and spread throughout Britain.[14] DNA evidence from bones from the islands reveals biological family relationships among the buried individuals.

Also found was a flat human figurine carved out of stone, with a round head atop a round body. The head has lines looking like a brow with eyes and a nose. On the body, a V-shaped line was etched, appearing to represent breasts. The Orkney Venus, also called the Westray Wife, was found in the ruins of a building, suggesting to archaeologists that it had been placed there during an act of closure.[15] Maybe placed when they were about to make their sudden departure.

The Ring of Brodgar, a mile away from the village, is aligned with the movements of the moon. It sits in a natural amphitheater of hills and on a high point along the spine of the isthmus. This ring of stone has a regular circular shape and once contained sixty stones, with only twenty-one remaining. The circle is surrounded by a ditch. There are two causeways, in an opposed arrangement, which seem to be points of entry and exit for some sort of procession. A second circle nearby, the Standing Stones of Stenness, is taller, its stones nineteen feet high and aligned with the movements of the sun and the moon. Both could have been ritual centers for the people living at Skara Brae. In the center of the ring is a large stone hearth, just like in the homes at Skara Brae. The landscapes on which the stones were set seem to have been specifically chosen to emphasize the most extreme rising and setting points of both the sun and the moon, aligning with where the solstice sun and the lunar solstice or standstill moon would rise.[16] The full moon appears low, large, and luminous at a point in an eighteen-and-a-half-year cycle. Called a standstill moon, it rises and skims the surface of the distant hills in the high latitude of the Orkney Islands before setting again.

Standing in the center of a ritually created stone cosmos while the stars, sun, and moon passed in circular paths above would be a moving experience. Surrounded by hills, it emphasizes the feeling that you, yourself, are at the center of the cosmos. For a people who may have believed their dead had traveled up into the sky, and become part of important solar and lunar events, the rings may have helped them honor the dead and other spirits.

Also nearby is another ancient dwelling site, the Links of Notland, containing homes built in a similar style. This site had a larger building set apart from the homes. Built into the wall of the building was a ring of cattle skulls, with the long horns interlocking, all facing inside the structure. Interestingly, the DNA from the cattle's bones show the people had bred their cattle with wild auroch.[17] This was one of the animals frequently painted on the walls of Ice Age caves, more than five thousand years before this structure was made.

By thirty-five hundred years ago, domesticated animal meat began to replace marine foraged protein. The human bones show changes in the isotope signatures. Increasing amounts of emmer and einkorn wheat, barley, peas, chickpeas, bitter vetch, lentils, wheat, and barley have been found in middens from this period. Domesticated cattle, goats, and pigs and also fermented milk in the form of butter or cheese were eaten. The grains and animals, once imported, became adapted to Europe.[18] As more cereal grains were grown and more livestock were raised, farm fields became more and more differentiated from the nearby natural environments. The alteration of nearby natural environments increased. Greater stretches of forest were cleared over time, and formerly natural meadows were altered to suit the farmers' needs. Fields were fertilized with dung from cattle, sheep, goats, and pigs and were irrigated.[19]

A NEW CULTURE moved in from Russia—the Yamnaya—about forty-five hundred years ago. They are alternatively referred to as the Corded Ware culture, based on pottery decorated by pressing string into wet clay. Their distinctive genes are detectable in modern Europeans.

Beginning around 2800 BCE, people began erecting massive mounds, "accenting the individuality of people, accenting the role of men, accenting weapons," says archaeologist Piotr Włodarczak.[20] Europeans' DNA contains mixtures of these ancient immigrants: Cro-Magnon from Africa, Anatolian farmers from the Middle East, and Yamnaya from the Russian steppe.

OVER TIME, many Neolithic stone structures were ransacked. Many stones were removed and reused to construct stone walls and to build people's homes. The Roman Catholic Church made use of the alignments to pray for spiritual intercession during a drought. Some Neolithic megaliths in Brittany were defaced by early Christians, who carved crosses into them. Lingering beliefs in the magic and sacredness of the stones probably helped to generate the respect that kept most megaliths preserved for so long. The stones were revered as living beings; legends say the stones had once been men who were changed into stone. Some French people still believe that Korrigans—fairies—live underneath the rocks and dance at night in circles. Even in the present, women will come and rub against certain stones, believing it will help them conceive. By the late 1890s, the Neolithic stones at Carnac were protected by the French government.

FEST-NOZ

Many modern Breton people feel strongly connected to the past. The evening after we visited Carnac, we attended a fest-noz—a modern musical reincarnation of an old harvest festival. Grain threshing was once done on the floor of a building or in the open, accompanied by music. Threshing was followed by a meal and cider drinking. Fest-noz is now held in civic centers and outdoors around Brittany as an exhibition of Breton cultural pride. We enter a huge room with raised bleachers on the side, facing a stage. In the center of the room, a large open space has been created for dancing. The back of the room is set up for eating and drinking cider, with booths selling cider, crêpes, and wine. Hundreds of people are gathered this night. A group of musicians begins playing bagpipes, Breton oboes—called bombards—accordions, drums, and guitars in a rhythmic tune. As the first notes ring out, nearly everyone, from the young to the old, quickly rise from their seats and move to the open space in the center to dance. Linking to one another with crooked arms and holding each other's little fingers, swinging arms, they dance a step-hop pattern, forming long intertwining lines.

We stand behind the lines of dancers, attempting to imitate their steps. A woman notices us there and demonstrates how to do the An Dro. We join in the back, linking up with a line of people, and clumsily dance, getting dragged along as the line moves around the room. Dancing continues for hours as different bands take over. Several different groups playing traditional music are followed by a band playing modern raplike Celtic songs, then another band playing traditional tunes. In between sets, people chat in Breton and French, admire children, and drink numerous cups of cider and beer. Worn out by our efforts to learn the dance steps, we sit back down and watch the Bretons dance together into the night.

HOW FARMING CHANGED

The beginning of farming shifted people's relationships to the land. Farming was once relatively benign, such as in agroforestry where crops are planted among existing trees and shrubs. Modern versions of ancient farming techniques, improving soil via agroecology, no-till biointensive farming, and localizing all food production, are now showcased as ways to help humanity increase our climate resilience and mitigate climate change.

Early farming used no-till methods, but tilling or plowing the soil became the dominant method when populations grew during the Roman Empire. By the seventeenth century, plows that cut deep furrows into the soil were used. As agriculture became more intensive, soil fertility decreased. Tilling exposed so much soil to the air, its reserve of stored carbon decayed. More fertilizers had to be applied and more land had to be brought into cultivation. More than a third of the earth's land surface has already been converted from wildlands to farm and grazing land. One million species of plants and animals, including pollinators, are endangered because of habitat loss to farms. According to the United Nations Intergovernmental Panel, land clearance for agriculture is the main cause of the land degradation driving today's species extinction crisis. As more and more habitat is converted to farmland, more streams are polluted with more fertilizer runoff and more topsoil washes away.[21]

Two hundred years ago, when farms were still small and remote compared with today, 90 percent of the U.S. population lived on small farms and grew their own food to eat. One hundred years ago, farms were still small, and over a third of the workforce were farmers. Farm families kept the land over generations. Farms supplied the local community with food. Farming was more sustainable when it was hand and horse powered.

Farming has intensified dramatically over the past one hundred years. Now, less than 2 percent of the U.S. population are farmers.[22] Most Americans farms are big, corporate farms that continue to plow up more and more grasslands. Farm mechanization began in the 1920s, powered by fossil fuels. Combines, machines combining harvesting and threshing, began to be used. Machines became self-propelled by the 1930s, not needing to be drawn by horse or tractor. Attachments were introduced to harvest corn and tomatoes, followed by mechanized seeders. By the 1970s, farmers were relying heavily on fossil fuels to power tractors and combine harvesters and supplying petroleum-derived fertilizers and pesticides to farm fields. These fertilizers are washed into streams and eventually to the ocean, nurturing tremendous algae blooms. The dying algae falls to the bottom, feeding bacteria and depleting oxygen, creating ocean dead zones.

Using fossil fuels indiscriminately became expensive for farmers. When, in 1973, then secretary of agriculture Earl Butz famously told farmers, "Get big or get out," many farmers had to get out. Those remaining consolidated their holdings. Mechanization led to larger fields and specialization, with one farm often raising a single crop.

On most corporate farms, undesirable animals and plants are killed directly by chemical pesticides. Neuro-active pesticides, neonicotinoids, have made agriculture even more toxic to insects. Neonic chemicals are a thousand times more toxic for bees than DDT and are one of the causes of decline in insect populations, especially affecting the health and reproductive abilities for honeybees and wild bees.[23] Both honeybees and wild native bees are responsible for pollinating a third of all human food crops. Fewer insects overall lead to fewer birds, since insects are important sources of food for birds.

Scientists at the Cedar Creek Long Term Ecological Research site have studied the long-term impact of conventional farming in Minnesota. They compared agricultural fields that had been abandoned between one and ninety-one years ago and then allowed to regrow as grasslands. When they compared the formerly plowed fields with other plots that had never been impacted by farming, plant recovery was slower in the former farm fields. Even ninety years later, the plowed fields had only three-quarters of the former plant diversity and half the plant productivity, which was measured using the weight of all the plants in a particular area.[24] Most disturbed ecosystems are unlikely to return to the undisturbed state they had before without intentional action to restore them, say the researchers. Even then, it might take a long time to return to their earlier more productive state.

Bird numbers have declined since 1970.[25] There are almost 30 percent fewer birds than there used to be—grassland birds have especially declined. "It's a wake-up call, we can work to ban harmful pesticides and fund effective bird conservation programs," says Michael Parr of the American Bird Conservancy. Waterfowl—ducks, geese, and swans—have rebounded over the past fifty years because of investments in conservation by hunters and by government funding for wetland protection and restoration. We can make similar investments for birds of the grasslands.

Wildlife-friendly farming, removing areas from production to provide wildlife habitat, is resurging in its different forms today. Wildlife corridors, hedgerows, restored riparian areas, native plant buffer strips, and orchards with understory plants for pollinators—all help conserve biodiversity. Farmer and writer Wendell Berry believes that today's farmers must return to having greater reverence for other living things. He advocates extending courtesy toward all animals to broaden a farmer's sympathies in their work.[26]

Regenerative agricultural practices, including organic farming, permaculture, and agroforestry, are based on old practices used by Indigenous people. These methods focus on restoring soil health to provide both human food and support a diversity of species and ecological functions. Soil contains diverse living components. Besides plants, there are one-celled bacteria, algae, fungi, and protozoa living in microscale environments between particles of soil, nematodes, tiny arthropods like oribatid mites, and earthworms and insects living in plant litter and humus. Together, these organisms decompose plant and animal material, store nutrients, make nitrogen available to use, enhance the porosity of the soil, and provide food for animals above ground.[27] By rebuilding the soil and minimizing its disturbance, farmers help decrease the need for artificial fertilizers, help soil increase its porosity to absorb water and stay in place, and grow nutritionally healthier food.

Soil stores the largest proportion of active carbon on earth. Conventional farming releases twelve thousand megatons of carbon dioxide to the atmosphere each year, contributing to climate change. It is estimated that one-third of global CO_2 emissions are contributed by the different stages of the conventional food production system, from the manufacture of fertilizer to disposable food packaging.[28] Worldwide, deforestation caused by increased agriculture accounts for nearly 20 percent of greenhouse gas emissions. But agroforestry, growing food under the protective canopy of the forest, reduces the need to clear land to plant crops. Regenerative farming practices, which restore soil health, build reservoirs of carbon within the soil. If these practices are enacted widely, soil processes could offset as much as one-third of global CO_2 emissions.[29]

By growing cover crops, rotating a diversity of plants, and significantly reducing pesticide use, farmers encourage beneficial soil life. Using cover crops and leaving the plant material to decay increases the amount of organic material added to the soil and helps the soil fungi and microbes to prosper. The hyphae of fungi grow in association with plant roots, growing out into the soil to absorb nutrients and transport them to the plant. Symbiotic microbes help break minerals down into substances usable by plants. The fungi make glomalin, which helps soil particles bind into aggregates, leaving pores in the soil for water and oxygen to move and soil biota to live.[30] The microbes living in the soil take more carbon from the air and turn it into plant tissue.

Organic farming practices fertilize crops using organic manure from livestock and using rotations of nitrogen-fixing plants. Crop rotation and intercropping can reduce agricultural pests without relying on pesticides. Mechanical weeding reduces the need to use herbicides. Farms can provide native habitat for bees and other pollinators to live, and habitat for beneficial pest-control organisms like green lacewings and parasitic wasps helps farmers reduce insect pests on crops. Farmers who remove a small area of land from production to re-create beneficial wildlife habitat did not lose much yield per hectare and could still earn a profit.[31] Even a few well-placed organic farms can create a spillover effect of plants needed by pollinators, for beneficial insects living near the ground, and for butterfly populations into neighboring conventionally farmed fields.[32] The demand for organic food has grown by 20 percent each year.

Writer and activist Wes Jackson recommends not tilling the soil altogether.[33] At the Land Institute in Kansas research sites he helped establish, ecologists experiment with farming methods that re-create the same rich, spongy topsoil, full of naturally occurring nutrients, once found in natural tallgrass prairie ecosystems. Like old-growth forests, some grasslands can be considered old-growth ecosystems, but indications of their age are hidden underground. The thick root system of perennial grasses stays alive though successive growing seasons; it holds the soil in place and soaks up water. A photo on the Land Institute's website shows a scientist holding an entire perennial grass plant over his head, the plant's root system cascading down like Rapunzel's golden hair, its ends curling along the floor.[34] These roots help absorb the broken-down metabolic products of organic matter created by the community of soil organisms as well as anchor the prairie soil. A new perennial grain, called Kernza, a type of perennial sunflower, perennial wheat, and perennial chickpeas have been developed by the Land Institute. These long-living plants supply food without much fertilizer, irrigation, or pesticides. Their harvesting results in less ground disturbance and erosion. Perennial plants help retain topsoil.

Hedgerows—once common in Europe, then plowed up to maximize farm production—are again being regenerated. Hedgerows helped prevent soil erosion and continue to benefit pollinators and provide corridors for birds and other wildlife to move along. British hedgerows contain woodland plants like bluebells and anemones, woody shrubs like hazel, dogwood, and guelder rose. Blackthorn, cherry, plum, elm, elder, hawthorn, and oak also grow in hedges. In different regions of Europe, hedgerows take on different characteristics.[35] Bocages in Brittany, pastureland divided into small hedged fields interspersed with groves of trees, help to control erosion by directing water flow on the land. Spanish hedgerows may consist primarily of ash trees. Hedges are important remnant habitat. Animals use them as a corridor from one field to the next, and the hedges are home to the birds and insects that consume insect pests like aphids. If hedgerows are replanted, existing hedgerows restored, and pesticides eliminated, many insect species can recover. Hedgerow habitat on farms can help protect pollinators like bees and butterflies.[36]

AGROECOLOGY IN PORTLAND

A transition to sustainable agriculture requires new knowledge—farmers networking with each other and with researchers. Farmers at the Mudbone Grown farm along the Columbia River floodplain near Portland, Oregon, share new information with each other regularly. Visiting the farm on a summer's day, we smelled the aroma of compost, the aroma of blooming flowers all around us, and saw for ourselves the dark and rich soil. Long rows of well-tended squash, peppers, cucumbers, tomatoes, and towering patches of sunflowers extend out a great distance. A field of sunflowers was planted to attract late-season bees.

Fewer young people are becoming farmers, so Mudbone started a young farmers incubator in the rich floodplain soil. They especially target young people of color interested in farming. Although farming goes way back into black culture, an 1844 Oregon law excluded black settlers, forbidding them from owning land. The laws remained on the books until 1926.[37] Flynn is one of nine people working here to become better farmers. Flynn grows medicinal herbs like valerian, elderberry, and chamomile. "I know how to grow these herbs that help calm anxiety, but I have never grown things on this scale before," she says as she stoops to pick a leaf of clary sage. She will use the sage to make an essential oil. "Farming isn't just a white-people thing," she remarks. She also grows flowers for sale. Patches of beautiful golden-red marigolds and colorful zinnias cluster around us. The plot of another young farmer, Serbie, has rows of bush beans and nicely mulched winter squash. "You use a broad fork to aerate the soil, and I add a lot of compost," she instructs us. "I want to farm so that I can feed myself and a few others." The produce grown in this farm goes to low-income families in the nearby community.

PERMACULTURE, one form of agroforestry, is another ambitious reform of conventional farming. It is referred to as practical ecology, where natural systems are re-created to build an ecological community that also produces food for people. Overstory trees are retained and new ones established, many of which are native to the local area, such as oaks. Locally native wildlife-friendly shrubs are planted to help support local wildlife species. These shrubs help create shade that lowers the temperature of both the air and the soil. The practice helps the structure of the soil maintain its carbon and basic nutrient-recycling capabilities. More water is also retained. Both cooler temperature and the presence of more water help young annual plant seedlings to grow. Permaculture has been called "edible restoration" because it yields food for people and helps restore native habitat for wildlife.[38]

At the Sherrett Food Forest, the use of permaculture eliminates the need to till the soil. A canopy of nut trees has nitrogen-fixing shrubs interplanted beneath. Annual vegetable plants are grown in intermediate places. Rainwater is stored in a catchment system and used for summer irrigation. Pollinator-supporting plants like cardoon, berries like gooseberry and currants, herbs like elderberry, and ground-cover plants are all integrated into the growing system. Meanwhile, in another area of the farm, staked tomatoes, rows of climbing beans, and beds full of kale grow lushly, alongside abundant weeds, as short-term cash crops for the farm as the founders continue to grow their forest habitat.

"It's hard for a farmer to make a living," admits Melissa, the manager of the Sherrett Food Forest. "We diversify our efforts. We have chef farm-to-table dinners and hold permaculture classes to raise more money. Growing food this way is not cheap." She enthusiastically describes a partnership recently developed with a Portland restaurant to use her produce in its dishes. This farm is one of more than fifty different community-supported agriculture (CSA) farms in the area. Households buy a share of the harvest; each week of the growing season, they get shares of vegetables, fruit, and eggs.

Layered agroforestry can increase soil carbon absorption by 2.8 tons of carbon per acre a year. Growing crops in an understory can help soil retain more carbon because the soil structure is largely maintained. There are over 250 million acres of natural tropical forests being maintained as agroforests to grow shade-grown coffee, chocolate, macadamia nuts, coconut, black pepper, pineapple, and bananas.[39]

Increasing local food production is another important solution to many ecological and social issues. In Portland, low-income families, many of whom are black and Latinx, lack access to healthy foods. White neighborhoods might have five times as many grocery stores as black neighborhoods, where there are many more fast-food restaurants than grocery stores. Fast-food restaurants sell cheap high-calorie foods instead of fresh foods. As a result of a diet of fast foods, poorer people tend to have high levels of obesity, diabetes, and hypertension. But local community gardens, where people can grow their own fresh produce, community-supported agriculture from local farms, and farmers markets placed into these communities all create a healthier trend. Food grown locally is generally fresher and higher in nutrients. Eating food grown within range means less water and less fuel is used to transport the food.

We visit one community garden in Portland's Cully neighborhood. Many different species of bees are actively foraging in oregano, sunflowers, and artichoke plants gone to flower. Bees fly among the tomatillos, cilantro, rows of beans, large zucchini plants, and rows of corn. The soil is rich and black with humus. "People have worked the soil here over the past thirty years," one gardener, Sandy, tells me. "No one uses pesticides here." Sandy, an energetic Hispanic woman, picks a yellow zucchini, pollinated by squash bees, and hands it to me. "I love this place—we have such a wonderful community of gardeners." We walk over to a luxuriant garden bed full of tomatoes, neat rows of potatoes, corn, lettuce, onions, peppers. "That's Carlos's bed, over there. Carlos fed his family with this garden over the years. His children have grown and left home." Sandy continues, "Now he donates most of his produce to the local food bank." In another plot, rows of tomatoes and round circles of squash vines are growing. "Isabella started by helping her mom garden. She is an adult now—she has her own plot."

Early on Sunday morning, we arrive at the farmers market in another Portland neighborhood. An array of stalls sell produce, berries, coffee, prepared meals like tamales, and African squash soup and okra stew. Heading over to the N & N Amaro Produce stall, which operates west of Portland near Gales Creek, we are greeted by the farmer, a smiling man sporting a black mustache. His wife is setting vegetables out on the tables. I stop in front of a pile of squash blossoms and ask him how I might prepare them. He explains how they might be steamed or fried, added to a quesadilla. He directs us over to an array of green herbs. I recognize the basil, the cilantro, and the parsley, but he points to another herb, epazote. "It's good for the digestion, to cook with the beans." He gives us some leaves from an herb that his wife uses to prepare the squash blossoms. A plant native to Central America, it is aromatic and has jagged green leaves. He breaks one off and offers it to Roger, who tastes it and says, "It's pretty good!" We purchase squash blossoms, the epazote. "Come back and tell me how it turns out," the farmer calls out, then adds, as we walk away, "but you must be careful to cut off the stems."

Growing and buying food locally supports the local economy. The farmers market organizers estimate that urban farms contribute over $1 billion in locally produced food that is sold directly to people like us at one of forty farmers markets, or to restaurants or food processing plants. There are many hops fields that sell their crop to breweries, including to the fifty-eight breweries in Portland. All totaled, the Oregon food economy—from local food production and produce prepared and served up to a diner or processed and sold as a product—had a value-added total of $28 billion in 2017.[40]

Agroecological methods, used over thousands of years, can increase local food production without harming soil and water. According to a UN study, sustainable food production doubled the crop yield as well as diversified the kinds of harvests for local communities.[41] Projects implemented in thirty-six countries over the past two decades show how biodiversity conservation can be integrated into agricultural practices.[42] These include developing wild crop relatives, pollinator-friendly practices, agroforestry, improvement of belowground soil biodiversity, community seed banks, and support for local farming livelihoods.

Later that day, we dine in a Portland restaurant where we are served glasses of Oregon pinot noir. Farmers and chefs Karl and Alix serve us plates of food they have grown on their Sauvie Island farm. Called Spätzle and Speck, their catering business uses the heirloom tomato and lettuce varieties they grow, fava beans, onions, garlic, snap peas, and other farm-grown vegetables. The farm has cherry and peach trees, and grows mint and more unusual herbs such as hepetela. Karl and Alex regularly go up to Mount Hood National Forest to forage for wild herbs, fiddlehead ferns in spring, and in the fall for morel mushrooms. Karl serves us a salad with tiny lettuce leaves, sprouted barley mint, and fat slices of pickled onion, tossed with a garlicky oil and sheep cheese. The hepetela tastes like oregano. Every bite of our dish has a slightly different delicious taste.

CHAPTER 4

From Picturesque to Ecological Preservation

WILDLANDS ARE PLACES where natural processes dominate and human touch is minimized. They are sources of clean water, help to clean air, and provide resilience to climate change. These places are reserves of biological diversity for many species of animals and plants, all of whom depend on the continuation of complex relationships. Wildlands are important places where modern people can restore their relationship with the natural world. But it wasn't always appreciated that way. Before Europeans and non-Indigenous Americans could find beauty in wild places, a turn to a new way of thinking was required. Romantic-era writers and poets of two centuries ago sought wildlands as places to encounter the sublime and be reminded of their infinite origins.

To better understand how Western society formed ideas to preserve land, we visit several wild places in Britain and the United States to hike.

LAKE DISTRICT NATIONAL PARK

We start walking England's Coast to Coast trail at St. Bees, following well-maintained stone walls, passing church spires and rolling green hills to reach the sea. The path heads north up and along the cliff, lush with vegetation, eventually turning inland. Eventually, our path runs up alongside a lively clear stream and through the pretty woods. A signpost points us upward, toward Dent Fell, a strenuous climb for us along a barren, steep, and muddy hill. A tall stone cairn sits at the summit. Roger and I turn back to look at where we have come from; the tidy English countryside below has patchwork green and brown fields lined by hedgerows. The blue Irish Sea shines beyond the fields. We're lucky with the weather; it is usually not this clear and calm here. We continue and enter the Lake District National Park.

This national park and other mountainous areas were seen by Europeans as wastelands until the end of the eighteenth century. Mountains were believed to be the devil's ruined and broken lair. Visitors passing through the Alps are said to have recoiled in terror, pulling down the window shades so as not to have to see the mountains. Walking in formal gardens was the acceptable outdoor physical activity. English settlers in North America brought this negative idea of wilderness with them. When the Puritans settled in New England, they set to work cutting down the vast forests, believing it was their God-ordained destiny to transform the forest into farms.[1] Early entrepreneurs set up companies to log New England and Virginia, selling logs and animal skins as commodities on the world market. Before long, so much forest was cleared that the land could not absorb the precipitation. There was widespread flooding and few game animals. The earliest environmental regulations in North America set aside trees and regulated hunting to compensate. But then a change occurred in Westerners' attitudes; England's Romantic philosophers began to write about nature as a place to encounter God. Western people began to seek out natural beauty. Wild landscapes were beautiful and sublime. The idea took hold in America in a philosophy called transcendentalism.

As we head down the opposite side of Dent Fell, we see a steep rounded hill, Raven Crag, rising before us. Our track takes us right up and over the top of the steep hill and, just as steeply, back down to the bottom where a wooded beck named Nannycatch Gate runs—a tiny, clear-running watercourse with a rocky bottom. Down along the streamside, we pass runners with numbers pinned to their backs, each breathing deeply as they slog along. We are happy to find the trail flattening out. After crossing the stream, we walk along the first of hundreds of miles of well-kept dry stone walls—no mortar is used—we will see in the days ahead.

Making a turn, we spot a spry gray-haired woman sitting on the side of the trail knitting an orange sweater while in a lawn chair. She is apparently the runners' cheerleader. As another group of runners passes her, she calls out "Nicely done!" We stop for a chat to learn about the run. Called the St. Begas Ultra Trail Race, it's a thirty-five-mile course from Ennerdale, where we are heading, to St. Bees, where we left this morning. She tells us this landscape is called a fell. "Fell running began long ago," she explains. "The runs were contests to identify the swiftest messenger for a town." She refers to a legend about King Canmore of Scotland who held a race in 1064 to determine the fastest runner to deliver his messages. As we start to walk again, she encourages us merrily in her bright voice, "Enjoy your walk!"

The next day, we travel along the shore of Ennerdale Water, and huge solitary oaks occasionally dapple us with their shade along the way. The lake is tranquil, and at its end is the rushing river, the Liza, flowing into the lake. We cross and begin a five-mile slog up a gravel road through a densely wooded area of plantation. Finally, the woods thin out, and we descend to a stream flowing beside a small, isolated cottage. The day has turned windy and cool. It's warm inside the cottage, though, and as we enter, there are already a dozen other hikers inside, gathered together in small groups. We overhear bits of conversation; two people complain about their own sore feet, and others gossip about the goings-on between a couple they both know. Pouring ourselves mugs of hot tea from the waiting kettle, we sit down and quickly eat the sandwiches from last night's lodging—thick slabs of cheddar cheese inside slices of brown bread. Then we devour the Mars bars also supplied. We arise to continue, following a narrow, steep track winding up the bare hills, following a little stream named Loft Beck. As we climb, we pass the first patches of blooming purple heather among the dry brown grasses.

The barrenness of the endless rolling hills is striking, and we're surprised to learn later how these hills were once covered in trees. Ice Age glaciers shaped the great crags we are approaching. After the glaciers receded, Britain developed extensive forests of birch and juniper. Evidence of their existence was left by tiny tree pollen grains deposited in lakebeds. These pollen grains were extracted in cores taken from the sediments and painstakingly identified by paleobotanists. Pollen grains have a resistant shell made from chiton that preserves it; its relative abundance in the lake cores shows if the plant was common or not. The birch and juniper pioneers were succeeded by forests of pine, with hazel growing in the understory. Over time, the soil developed into thicker layers. Oaks began to grow in well-drained soil, and elm and ash forests grew in lowland areas.[2]

WHEN THE FIRST Neolithic settlers arrived in England five thousand years ago, they encountered a forested country. Gradually, they cleared the forest for farming and grazing land, felling the elm, the ash, the oak, and finally clearing the forest completely in many places. When contemporary soil scientists take soil cores, they find a decrease in oak pollen and an increase in cereal grain pollen. Megalithic stone circles were erected around the same time. Some woodland areas, like at High Furness, were not cleared and were subsequently managed for centuries as a source of wood. The inhabitants used coppicing, a traditional way of managing woods by periodically cutting the young shoots of big trees for wood. Cutting wood stimulated the main trunks to sprout new stems; a tree could live this way for a long time.

By the time William the Conqueror arrived in England in 1066, only 15 percent of the original forest was left. During the ensuing millennium, forests continued to be felled for wood to build England's navy. Only a few trees survive here now. Sheep are everywhere, cropping the grass, eating any potential tree seedlings, and leaving messy piles of dung everywhere. England is one of Europe's least wooded countries, though woodlands in Scotland have expanded through tree-planting programs. Forests are increasingly appreciated as places to hike and to appreciate wildlife.

We climb to the peak at a great round how called Haystacks ("how" being the local name for a top). The ridgetops are covered in heather. We are past the blooming season when there would have been a deep purple carpet everywhere, but patches are still in flower. Glorious views surround us. We turn back to look out over three different river valleys in the distance, each filled with a ribbon lake: Ennerdale Water, Buttermere, and Crummock Water. They shine brightly in the occasional sun breaks, stretching out to the north and west. The valleys were once glacier-carved, and as glaciers receded, they filled with water. Dammed by glacial moraine, the meltwater remained as lakes, now continually replenished by rain and streamflow. Other former ribbon lakes were filled with sediment over time, becoming river valleys, then farms.

The number of other people passing us on the trail steadily increases. We see why when we spot a parking lot full of tour buses and many cars down below. The Lake District National Park is very popular. More than fifteen million visitors came to the Lake District the previous year.

AT OUR GUEST HOME that night, we meet an older Englishman, Jeffrey, who looks just like the actor James Cromwell, the farmer in the movie *Babe*. He becomes more endearing in an old-grandad-font-of-wisdom sort of way as we talk. Jeffrey has taken many long hikes, but he enjoys the most those when he is accompanied by his grandson. He describes the lake landscape as opening out before you "slowly and gently unfolding as you walk along." His description rings true over the next few days. As we continue along the trail, I notice how different these sculpted rounded hills are from the rugged mountains of Oregon. This place has been tamed over centuries.

The next morning as we pass a cluster of cozy stone cottages standing along a beck, I see what has altered these hillslopes. When this little stream meets up with a larger stream, Greenup Gill, to form a torrent of white water, more sheep appear. When we follow this path as it continues to rise, even more sheep graze on the hillslopes. Splotched with different colors to mark to whom they belong, they are raised for either meat or breeding stock.

Fog builds as we continue to climb and reach a high boggy stretch. We find the path that leads along the crest of the grass-covered rolling hills. As the fog clears, we see the distant blue hills beyond today's destination—the distant grassy slopes looking soft to the touch, like velour. Downhill we go along Wythburn Valley, following along burbling Easedale Beck before finally reaching the small town of Grasmere. The best known of the Romantic-era poets lived here in a cottage two hundred years ago. Wordsworth was inspired about leading a better life by his walks through the Lake District.

"WE HAVE GIVEN our hearts away...we are out of tune," wrote Wordsworth in his poem "The World Is Too Much with Us," referring to the many problems created by the industrialized society of his day. The skies of the factory towns in England were full of smoke and filth from burning coal. More working-class people were forced to move to factory towns. They had been denied access to open fields to grow their own food when, between 1750 and 1860, great expanses of land were "enclosed," the open lands turned into private property. Many became laborers on large estates, supplementing their income at home by making fiber and weaving cloth. But the factories soon outpriced them with cheaper cloth and fiber. These working people could no longer support themselves. Entire country villages were abandoned as people moved to the cities. Urban rivers became choked with sewage. Factory workers were sickly; widespread epidemics of cholera, tuberculosis, and typhoid killed tens of thousands of people.

Wordsworth's vision of nature contrasted with those dire urban conditions. The poet wrote about the sublime feelings he experienced when out in nature. Wordsworth believed nature's power could enliven the downtrodden spirit and help rescue people from the working class. His poems encouraged others to become aware of the outer wild and their own inner wildness. Animistic and pagan ideas from ancient Greece, Buddhism, and Hinduism about everything being sacred were being reintroduced to Europe at the time. People could encounter the sublime if they would go out and walk in the wild. Once out there, they could experience nature's awe and beauty. In doing so, they would encounter the infinite God and be spiritually transformed and healed. The boundless energy of the sublime underpinned all of nature, Wordsworth believed.[3]

His vision inspired the creation of the first public parks in England's industrial towns, places for people other than the wealthy classes to go. Parks in Europe had been exclusive royal hunting grounds and private gardens for the rich. This changed in the early nineteenth century, when France turned former royal hunting parks, such as the Bois de Boulogne and Bois de Vincennes in Paris, into public parks. In England, the first public park was Birkenhead near Liverpool, followed by new parks created in London and Edinburgh. The scenery was re-created so as to inspire the sublime feelings of awe, placing rocks just so to form waterfalls, all intended to rebuild workers' health.

Wordsworth and his colleagues formed a European literary and philosophical movement stressing a love of being in nature and encouraging people to become more receptive to it. Landscape paintings of the time began to feature some quasi-religious elements, emphasizing the "sublime" and picturesque elements of nature. J. M. Turner and other artists emphasized the picturesque qualities of Lake District vistas on their canvases. Fearsome craggy mountains and streaming rays of sunlight showed the thrilling danger and picturesque quality of the landscape.

In America, Ralph Waldo Emerson developed his own offshoot of Romanticism called transcendentalism. Emerson met William

Wordsworth in 1833 while visiting England. Emerson had been a Unitarian minister, but grew to reject organized religion. He believed people were essentially pure and could achieve a deep spiritual awareness by immersing themselves in nature. In this way, everyone could understand the unity of all creation. He wrote, "Man needs no church to connect to the divine in the woods."[4] Inspired by Emerson, Henry David Thoreau described in his books the vital life force contained in rocks, ponds, and mountains. Thoreau inspired generations of people to walk in picturesque American landscapes. John Muir, who read Thoreau avidly, encouraged others to walk in the mountains and to revere the trees. He wrote about becoming inspired to preserve the landscape to enjoy nature's sublime qualities. Muir is remembered for arousing America's preservation movement and inspiring the creation of our national parks.

More people moved to cities in the late nineteenth century to work in factories. Correspondingly, an appreciation of the redemptive aspect of nature grew. Roads became more passable, options for public transportation increased, and more people could avoid walking long distances out of necessity. People could take streetcars, omnibuses, and trains to get out of the city. Middle-class people began heading to natural areas to walk for pleasure.

American cities developed public parks with paths for walking, and built urban playgrounds intended to improve the health of city people through exercise. Canal towpaths provided places to walk alongside the water. Hiking clubs like the Appalachian Mountain Club in New Hampshire and the Sierra Club in California were formed to organize group outings. An American culture of hiking developed—a culture that emphasizes the health and spiritual impacts for those who go out and take walks. During the twentieth century, improvements in hiking equipment and a new system of interstate highways made it easier for more people to visit remote areas to hike and camp.

LEAVING GRASMERE, we turn to walk along a path following Tongue Gill, heading uphill. Our trail zigzags up and up and then down and down to the bucolic town of Grisedale, where we follow the tree-lined Grisedale Beck. Remarkably muscular-looking tree roots skirt the banks. We pass farms, more pine plantations, continuous stone walls, and many more flocks of sheep. A platoon of hunting dogs brays in the distance. At the end of a stone wall Roger and I turn and walk under spreading oak trees toward the lake of Ullswater.

We spend the night at Noran Bank Farm. Heather, the proprietor, is the daughter of a shepherd in a long line of shepherds. A serious woman, she tells us about Beatrix Potter—author of the Peter Rabbit books—who had lived nearby. Potter was nostalgic for the old days when shepherds walked along with their flocks. When she bequeathed four thousand acres of land to the National Trust upon her death, she ensured that sheep farming would be maintained.

Heather and her husband work to maintain the stone walls, the fields, and her sheep. She believes that fell farming and shepherding are customary use of this land. She says shepherds have earned their right over centuries to have priority over other uses of land. She explains, "We believe the land is here for us to make a living." Though her livelihood now depends on tourism, she seems unaffected by Wordsworth.

Today's shepherds don't walk the land with their shepherd's crook; they use gasoline quad bikes to herd sheep. There are millions of sheep in England who collectively damage the land—to the detriment of vegetation, wildlife, and bird life. Grazing compacts and destabilizes the soil, reducing its ability to absorb water. During rainstorms, the runoff down the hills in sheets causes erosion. Sheep eat the young trees and shrubs that would have helped stabilize the soil, absorb the rain, and provide habitat for other species. Sheep make the soil boggy; their urine turns it acidic. Reducing sheep numbers, especially here in the park, could help halt the decline of biodiversity.[5]

Fencing the sheep out, along with active restoration, can rewild Lakeland.[6] Rewilding means restoring natural processes that allow a landscape to return to a more ecologically diverse condition. By reintroducing species, the top-down interactions of an ecosystem are restored in what is called trophic rewilding. Grazing animals that cause ecological harm, such as sheep, are removed and replaced with grazers that are closely related to those animals that were once here but are now extinct, such as Galloway cattle. After the grazers' reintroduction, their predators—hawks, eagles, lynx, and wolves—soon appear. Eventually, the system can become more self-regulating and not require too much management. Nearby, ecologists working in the Wild Ennerdale project experimented with reducing the number of sheep and replacing them with Galloway cattle. A once overgrown conifer forest became more diverse, with native broadleaf trees and open valleys. The marsh fritillary butterfly and beaver were reintroduced, and populations of voles and deer soon appeared. Elsewhere in Britain, woodlands have expanded and some areas are being coppiced again. This process of cutting stems from resprouting deciduous trees will temporarily open up patches for roe deer, badgers, stag beetles, and butterflies. The increased sunlight benefits bluebells. Their purple-blue fragrant flowers carpet the tree gaps on the forest floor in spring.

No DOMESTIC SHEEP graze in U.S. national parks. Yosemite, one of the first, was set aside over concern about sheep overgrazing the high meadows. Early on, national park managers believed they were preserving places as if the land was unchanged from a time before the European conquest of America. When Yellowstone became the first official national park in 1872, it was envisioned to "preserve the scenery and natural objects as a pleasuring ground."[7] Some park advocates envisioned Native people continuing their way of life within some national parks, but most Euro-Americans wanted a people-free wild nature in unique places like Yellowstone or Yosemite. Native Americans were forced off of parklands to lure white visitors, who were thought to be fearful of the natives.[8] Fresh in the minds of park planners was the recent history of U.S. cavalry forcibly displacing many Native tribes, sometimes murdering entire villages, which had been followed by reprisal attacks. Native Americans were not included in the park landscape to preserve the illusion of an unpeopled landscape.

Park managers did not recognize at the time how these lands had been managed by Native Americans, who over thousands of years had used low-intensity burning to maintain a mosaic of open prairies for hunting and "gardens" for basketry material and their preferred foods: camas, acorns, hazelnuts, and mountain huckleberry. Native people on the East Coast had been decimated by diseases brought by the first explorers and trappers. When European settlers arrived, the forests were thickly overgrown in unbroken expanses of hemlock, sugar maple, American beech, red oak, American chestnut, and sweet birch. Early explorers in the Pacific Northwest encountered the vast parklike open forests of ponderosa pine. Native people in the Pacific Northwest did not succumb to diseases until later and maintained their practices of periodic burning into the early 1800s.

IN THE MORNING, we walk across Patterdale Commons and follow a winding trail up into the bracken-covered hills. The spreading countryside below gleams, each field neatly bordered with its green hedgerows. Turning back again to the trail, we walk uphill and pass a small glacial lake, called a tarn, sitting in a bowl-like depression and surrounded by grassy rounded hills. We reach a high and lonely volcanic remnant named Kidsty Pike. Standing high up in a strong wind, I look a long way down and see our path continuing steeply down along grassy and heather-covered hills to a large body of water. Once we reach the bottom, we walk through a lakeside forest along the reservoir. Finally, the forest thins out, and we approach a row of tidy houses along the far bank of the lake.

The next day is cloudy as we continue along a pleasant stream. Large oaks are dispersed throughout the rolling landscape. Our trail eventually crosses the River Lowther, going up and over a perfect arching stone bridge. We pass the stone ruin of Shap Abbey and traverse fields of pink willow herb and buzzing bees. Once we cross over the pedestrian footbridge of the busy M6 motorway, the path enters a barren, flat country. We have left the lakes and entered the moors—a landscape also created by human activity.

The moors are dreary, dripping, and gray when viewed from the top of Ravensworth Fell. Rain begins in earnest, and the boggy soil squelches under our shoes as we trudge onward. Rumbles of thunder hurry us along. The trail leads us down to a wide, incredibly boggy area with a deep stream. We bend down and to take off our shoes and socks, crossing the cold, swollen beck. The water foams over the rocks, making a brownish swirl. Reshoeing ourselves on the far bank, we continue along the trail through the rain. Eventually, passing stone walls, we reach the outskirts of Orton and the endpoint of our walk. We have hiked over eighty miles.

THE UNITED KINGDOM FAR OUTDOES the United States in providing footpaths for walking. People who love to walk can choose from over 140,000 miles of footpaths in England and Wales. Some of these trails have sections that have been walked for centuries as people took their goods to markets to sell. France has over 180,000 kilometers of walking trails that crisscross the country, many of which pass through private lands. France's network of long-distance footpaths—the Grand Randonnée—can take a walker from the Atlantic Ocean all the way to Nice on the Mediterranean's Côte d'Azur, or along the pink granite coast of Brittany, or on a circuit around Mont Blanc. In England in 1932, four hundred people defied the enclosure law and walked across the moors of Kinder Scout in the Peak District. England and other European countries now have "right to roam" laws legalizing people's right to walk on areas of open upland and common land as well as on footpaths.[9]

Americans had access to large tracts of land for hunting and fishing until the Civil War. Afterward, landowners strung up barbed wire to close off large expanses of land. With the exception of several classic American trails, like the Pacific Crest Trail and the Appalachian Trail, there are many fewer hiking trails in the United States than in Europe. Americans can't walk freely across the countryside, through forests, crossing streams and meadows. American hikers are forbidden to enter private lands; it is viewed as illegal trespassing.

People are changed by taking long walks. Andrew Forsthoefel walked across the United States, taking backroads all the way from his home in Pennsylvania to Oregon. He deliberately chose quiet rural roads to walk on so he could encounter and interact with people and listen to their stories. He describes what happened to him as "life changing." His encounters with people he never would have met otherwise challenged many of his preconceived notions. Forsthoefel frequently experienced strangers' bottomless generosity and kindness.[10] Bill LeBon crossed the United States on rollerblades. He rollerbladed all the way from California to Washington, DC, stopping along the way to talk with anyone he met who would listen about climate change. Once he reached DC, he met with congressional representatives to talk about climate change. His experiences changed him from an angry young man to an activist who strives to find commonality with people he used to oppose.

Wildlands are important places where modern people can spend time outdoors. More people live in cities with less contact with nature than at any time before. Many feel increasingly estranged from nature, even as we continue to discover how being out in nature helps us stay healthy, physically and psychologically. Time spent in nature is well known to reduce stress, increase one's sensory awareness, heal attention deficit disorder, improve mental health, and increase one's ability to cope.[11] According to a recent study, even only two hours a week outdoors can boost someone's feeling of health and well-being. Researcher Matthew White explained that this health benefit of spending time outdoors holds true for everyone, independent of their ethnicity, age, income, and "even for those living with long-term illnesses."[12]

AMERICAN NATIONAL PARKS

National parks were an American invention that grew as wildlands continued to succumb to development. Yosemite Valley and the Mariposa Grove of Giant Sequoias were the first to be set aside in 1864. Forty years later, the national parks system was formed to preserve Yosemite and other examples of magnificent wild landscapes for the public. The U.S. national parks system has sixty-two parks, and over 125 national monuments.

John Muir is credited with inspiring the national parks system in the United States. Muir had carefully read Henry David Thoreau's work and also believed nature was infused with a divine force, a force that could help redeem "man." Muir had his own mystical encounters with the sublime while out in the mountains of California. Muir's strict Christian upbringing, his study of the natural sciences, and his observations of the destruction of America's wildlands combined to form his philosophy.

Like Thoreau, Muir was influenced by the beliefs of America's Native peoples. He wrote extensively about California Indians, Tlingit, and Eskimo people of Alaska. Although his relationship with Indigenous people was problematic—he made derogatory remarks about the Maidu—what he had learned after spending time with Native people shaped his own connection with nature. It opened him to even greater reverence.[13] He polished his nature theology, writing that God was the original conservationist: "Any fool can destroy trees. . . God has cared for these trees, saved them from drought, disease, avalanches, and a thousand straining, leveling tempests and floods; but he cannot save them from fools."[14]

Writing in the first person, using "I" or "we," and with an engaging tone, Muir charmed readers to empathize with and advocate for the preservation of wilderness. Wild nature was what we all needed to set ourselves free and appreciate the godlike "universal light and beauty" of nature. He equated being in the wild to an encounter with God. His books and magazine articles promoted more widespread wilderness appreciation.

Muir described his vision of preservation and his goal of saving all of Yosemite from development around a campfire with President Theodore Roosevelt in the Sierras. But preservation did not always go the way Muir envisioned. Later, Muir would disparage the government's decision to dam Hetch Hetchy Valley, the valley next to Yosemite and equally as beautiful, for drinking water for San Francisco. "Dam Hetch Hetchy?" he exclaimed; "[Might] as well dam for water tanks the cathedrals and churches!"[15] Muir died soon afterward, some say from a broken heart.

The Sierra Club was inaugurated by John Muir as a hiking club, but soon began to influence environmental policy. The Sierra Club led an anti-dam campaign against the proposed Echo Park Dam intended to be built right in the Grand Canyon in 1953. The proposers of the dam, the U.S. Bureau of Reclamation, absurdly claimed that the lake that would form behind the dam would enable people to get close to the walls of the canyon and see it up close. The Sierra Club director David Brower echoed John Muir in a mocking statement. "Dam the Grand Canyon? Should we flood the Sistine Chapel so tourists can get nearer the ceiling?"[16] The Echo Park Dam project was cancelled, but the Bureau built its dam downstream anyway.

Muir was instrumental in preserving Yosemite National Park. Several decades ago, I visited the park to camp and hike. There were a moderate number of people on the trail the next morning as we headed up to Half Dome, the iconic and brilliant granite dome. After several hours of scrambling up, we stood together at the summit, gazing out around us and down into the valley as if we were birds of prey. These days, Yosemite is visited by millions annually, by four and a half million in 2019. The better-known trails in the park are overcrowded. Roads are so overcrowded that cars can only crawl along with three-hour delays to get there. On arriving at a parking lot, visitors then compete for a parking spot.[17] If John Muir had prevailed and saved Hetch Hetchy Valley from being drowned in a reservoir, there would not be such horrific overcrowding in Yosemite Valley today. If Hetch Hetchy Valley had been preserved, visitors would be hiking and camping there as well, basking in its light and beauty.

NATIONAL PARKS and other federally preserved lands like national forests and national wildlife refuges are critical for species preservation. But parks cannot preserve species if they exist as islands within areas of degraded habitat. Many species of large animals migrate, many need large areas of habitat to prosper, and all need access to a large population to prevent inbreeding. The smaller the population, the less adapted it becomes over time to changing conditions, leading to extinction. Connectivity between parks and other areas is vital.

Underway around Yellowstone National Park is a large-scale plan, coordinating efforts between public and private interests, to restore degraded areas of habitat and interlink them with already preserved areas. Proponents dubbed this effort "Y2Y"—Yukon to Yellowstone—and developed sets of agreements with landowners and agencies managing the land throughout this vast area.[18] Significant parts of the region are still intact. Twenty million hectares of over 100 million hectares are now protected, and another 30.9 million hectares have some form of lesser protection.

THAT PARKS and other public lands continue to exist is never ensured. A constituency of people who love and actively participate in keeping them protected is required. Over the years, this constituency has been mostly white and well off. Those who support parks frequently remember having visited these places as children with their families. But what about those who did not get to visit parks as kids? Who visits, and consequently who cares about parks, is shaped by the history of discrimination of Native peoples and black people. Native Americans were removed when Yosemite, Grand Canyon, and Glacier National Parks were created. Black people were once excluded from visiting parks, and many say they still don't feel welcome. Some black people mention how uncomfortable they feel while hiking, worrying if someone might call the police, or if an interaction with another hiker will be antagonistic.

Portraying the Indigenous and black heritage in the parks is one of many changes undertaken to help formerly marginalized groups feel more comfortable visiting. Stories about the black army cavalry unit called Buffalo Soldiers, who helped fight wildfires and stop poaching in Yosemite National Park, are now told to visitors.[19] Stories about the Native American heritage at each park is included as more Indigenous people work in the parks. Who will wear the uniforms in the future if young people of color can't envision themselves working in parks or other natural places? Having more park rangers of diverse background might help change perceptions about whom the parks are for. Bringing park naturalist programs to the city, or bringing city kids out to visit a park through programs like Every Kid in the Park, are other efforts to increase interest in national parks.

Some people may not want to hike when they arrive in a park; they might prefer to spend time with their family and friends around a campfire or sit together by a lake to enjoy the view. By accommodating different preferences for outdoor activities, more groups of people may feel welcomed to national parks. Some potential visitors may not be able to access the park easily. Providing public transportation from cities to nearby parks is one way to accommodate groups of underserved people.

THE GRAND CANYON

I love our national parks; I have camped in and hiked around many of them. I've gone wild swimming in their rivers and lakes. Roger and I spent our honeymoon backpacking through the furrowed wind- and rain-carved hoodoo spires of Canyonlands National Park. Every ranger-led campfire program I have attended, from the Everglades to Denali National Park, has been well presented and meaningful. I worked as a national park naturalist for four summers, leading people on walks and explaining the nature of the forest ecosystem. I donate money to help support the parks. But I had never visited the Grand Canyon. During a recent spring, we traveled to Arizona to visit this iconic park.

ROGER AND I arrive and immediately walk over to stand at the canyon's rim; we gaze down seven thousand feet to the Colorado River. Standing nearby is a Park Service ranger. He tells us this is the second most-visited national park in the United States. Nearly six million people came to the Grand Canyon the prior year. We see throngs of visitors wandering among the gift shops, restaurants, and hotels at the rim. They may stroll along the trail that follows the canyon rim, and may venture a hundred feet or so down a trail before turning back. The ranger explains: "Most only stay long enough to snap a photo of the canyon. Then they'll move on to Zion, Bryce, and Arches National Parks." We watch groups of young people pose to snap selfies with the view of the canyon behind them.

Beyond the rim, the canyon is almost entirely wilderness. We seek a more transcendent experience and will hike down into the canyon tomorrow. "We are a wild species," wrote Wallace Stegner, explaining further how urban Americans have become sick, embittered, and desperate. Stegner coined an often-used phrase about the importance of national parks in a letter to Congress: "We simply need that wild country. . .as a part of the geography of hope."[20]

What do we hope for? Perhaps a recovery of long-lost senses that tie us to the natural world. Or that humans will mature into a more thoughtful and compassionate species. Hope that we might treasure the myriad other species who exist alongside ours and take their welfare into account when contemplating changes to the land.

When the Grand Canyon was set aside as a national park, the philosophy of park management had morphed into providing "for the enjoyment [of the scenery and the natural and historic objects and the wild life therein]. . .as will leave them unimpaired for the enjoyment of future generations."[21] Predators including mountain lions were shot in the Grand Canyon in the 1920s in the belief that this action would lead to more deer to please tourists. After the predators were killed, the deer population erupted. A young wildlife biologist named Aldo Leopold was distressed by this procedure and demonstrated how the lack of predators leads to an overpopulation of deer, degrading the Kaibab Plateau habitat by overgrazing.[22] Leopold went on to become a preeminent philosopher of conservation. Although he did not reference any Native Americans, his "land ethic" seems like a direct Western interpretation of a Native American worldview.[23] "When we see land as a community to which we belong, we. . .use it with love and respect," he wrote in his environmental classic, *A Sand County Almanac.*[24]

Predators were also shot in Yellowstone National Park—the last wolf was killed in the 1930s. Afterward, the elk population exploded and many became diseased. The elk overgrazed the aspen trees along stream corridors, causing erosion and thermal distress for wild fish. Wolves were reintroduced to Yellowstone in 1996. Shortly afterward, the elk population was halved; bison, deer, antelope, and other herbivores were unaffected. The role of wolves in generating greater ecological complexity has been recognized. The carcasses they leave feed many other animals, from grizzly bears to eagles. This ecological phenomenon is called "top-down" control. Since being reintroduced, wolves have been surprisingly adaptable, but enmity for wolves persists among some cattle ranchers.

THE NEXT MORNING we stand again on the rim and look down on the Colorado River before beginning to hike. Here, the Colorado Plateau continues its slow uplift as the Pacific tectonic plate thrusts itself underneath North America. The Colorado River down below has carved itself down into the flat-lying rocks. The place seems eternal, as if the canyon has always been this way. But the earth is in flux. Slow change over time, the erosion and deposition of sediments and their consolidation into rock, the earth's moving plates forming ocean basins and mountain chains, coupled with brief catastrophic volcanic activity, earthquakes, and intense flooding explain the origins of landforms before us.

The switchbacks at the top of the South Kaibab Trail are icy in the morning, so we strap crampons to our boots to grip the slippery surfaces. Although it's a mile down to the floor of the canyon as the raven flies, the trail zigs and zags for 6.5 miles to the river. Steep steps down followed by a steep downward-ramping trail take us past layers of limestone. The trail follows alongside a cliff, and eventually we round a turn to confront our first vista, Ooh Aah Point. We step out to a rocky ledge alongside a man from New York and a couple from Sweden. We stand together in silent awe and feel an immediate kinship before the deep gash of brightly colored strata of the Grand Canyon.

We continue down, passing layers of different-colored sandstones: red, white, and tan-colored layers of sedimentary rock sandwiched on top of one another. Sandstone provides distinct resistant rock caps on the mesas in the vista before us. Below each distant mesa are greenish aprons of eroded debris from a long-gone mudstone layer. Now the trail turns out from the canyon walls and heads into the abyss below along a series of ridges rising above the depths. We walk along several steeply descending stretches and reach a second viewpoint, Cedar Ridge. Roger snaps a photo of the long ribbon of the Tonto Trail that stretches and winds across the miles. The musical descending notes of a canyon wren, *tewee tewee tewee,* cheers us onward.

We have descended two thousand feet by the time we reach Skeleton Point. Looking down into the canyon from here, I see the dark basement rock at the bottom of the canyon. Below the dark rock flows the chocolate brown Colorado River. The Vishnu metamorphic rock at the bottom is the oldest visible rock in the canyon. Named after the Hindu god of creation, Vishnu, this rock is part of the earth's skeleton. It is in fact the core of the North American continent, the older foundation of the earth's crust—the craton. Here is one of the few places where we can see the craton ourselves. Ninety miles thick in some places, it rides on top of the solid yet fluid layer of the earth's mantle. Many other rocks have been deposited, accreted on from elsewhere, or extruded from magma onto the North American tectonic plate.

Although the Vishnu rock is a whopping 1.8 billion years old, our world is considerably older. If we imaginatively compress the 4.6 billion years believed to be the age of the earth into a single year, the Vishnu rocks formed in the middle of the metaphorical year. It began as a granite pluton and was transformed by high temperatures and pressure into dark gneiss and schists.[25]

Each successive rock layer in the canyon is a remnant of an ancient environment. Lying above the Vishnu Formation is a reddish-brown Precambrian limestone, deposited in an ancient ocean in late July of our metaphorical year. This rock bears stromatolite fossils from ancient blue-green algae, one billion years old. This species of algae survives today off the coast of Australia.

Traveling up from the Vishnu Formation through the limestone, we come to a big gap, called an unconformity, between the limestone layer and the sandstone layer above it—the rock that was once there has eroded away. We cannot observe fossils from the Ediacaran Period in the Grand Canyon, the time when animals were soft-bodied yet moved around the sea floor with no obvious defenses. Nor can we observe the period when predators emerged and the little soft-bodied animals disappeared. Animals evolved armor or skeletal shells as protection, while other animals burrowed into the sediments beneath biological surface mats as a refuge.[26]

The next visible layer is a red sandstone, formed in an ancient ocean in our metaphorical late November. It contains trilobite fossils—a segmented marine arthropod—and clamlike brachiopods, as well as fossil jellyfish. In the next layer of red Supai rocks, a succession of shale, sandstones, and limestones were deposited one on top of the other, along with amphibian fossils. The Kaibab Formation lies above this, a yellowish limestone layer mixed with shallow ocean shales, sea "lily" fossils (crinoids), and echinoderms, animals related to sea urchins and sea stars. Sea lilies still live in the ocean today, at depths of a thousand feet. A thick Redwall Limestone layer bearing fish fossils lies on top, followed by a layer containing land plant fossils from ancient lush tropical forests. All the continents fit together at this time in the supercontinent Pangaea.

A series of massive volcanic eruptions and massive basalt lava flows from the earth's mantle broke Pangaea apart. The Atlantic Ocean opened up as the sea-floor spread along the Mid-Atlantic Ridge. The lava emitted sulfur dioxide, causing acid rain, and led to a 15°F rise in temperature. The great Permian extinction ensued; 96 percent of all ocean species and 70 percent of all terrestrial vertebrate species vanished. The Mid-Atlantic Ridge continues to push the North American continent west at an average rate that matches the growth of our fingernails, an inch a year.

The Mesozoic layer is not visible to us here in the canyon, but can be seen nearby at the Vermilion Cliffs. Dinosaurs that flew, swam, and walked prospered. Small mammals, birds, and plants with flowers appeared. The Chicxulub "impactor" plummeted into the atmosphere and crashed into the Caribbean Sea off the coast of Mexico. Seismic shaking from the impact triggered massive basalt flows, again accompanied by sulfur dioxide emissions, resulting in another ecological catastrophe. Half of all life forms became extinct. The present era, the age of mammals, began relatively recently. The Kaibab Plateau began to rise about thirty-five million years ago, beginning the creation of this canyon.

Still imagining the age of the earth compressed into the timeline of a year, *Homo sapiens* evolved at 11:48 p.m. on the last day of the year. All of recorded human history, as significant as we find it, fits into the last second of time.

Roger and I ascend the trail from Skeleton Point to the rim. Later that night, we step out and stand again at the canyon's rim. It takes some minutes before our eyes adjust. This is a "dark sky" park; the stars of the Milky Way clearly appear before us, extending in a slight arc from the east up to the western sky. We imagine we can see a distinct bulge at the center of the galaxy. There, 26,000 light years away in the center of our rotating, spiraling, pinwheel galaxy, resides a massive black hole surrounded by about one billion stars. Roger spots the distant Andromeda Galaxy, home to about another trillion stars, about 2.5 million light years away. Other distant stars might be other galaxies. Astronomers estimate the universe has two trillion galaxies, each with billions of stars. These are either visible or inferred using today's sophisticated telescopes.[27]

In the modern story of how we came to be, humans arose out of the same processes that formed the stars and this planet. Hydrogen and helium were products of the Big Bang. Other elements formed within stars; carbon and oxygen atoms formed from stellar nucleosynthesis. Gases and cosmic dust from the solar system's creation coalesced to form Earth; heavier elements like iron and nickel form the core, while lighter silica and oxygen rise to the top to form the crust. Carbon molecules called biogenic graphite, indicating early life, were discovered in Labrador metasedimentary rocks nearly 4 billion years old.[28] Using a combination of fossil evidence and DNA codes common to all species of life, scientists have calculated that the earliest common ancestor of all life might have originated as far back as 3.9 billion years ago.[29]

We are the product of the billions of years of Earth's history in whom the earth reflects about itself. Buddhist Alan Watts reminds us that humans literally came out of this world. "You are not a stranger here," he writes, comfortingly.[30]

CHAPTER 5

Down the Oregon Coast Trail

OREGON'S COASTLINE is renowned for its beauty. It's possible to find beaches perfect for solitude, where no one else is out walking. Waves crash on a rocky headland, and sea lions bark from offshore remote sea stacks as tumbling storm clouds cover and uncover the sun. Massive geologic forces shaped this coast. Some are barely perceivable, like the way the beach emerges upward by an inch every five years as the Pacific Plate shoves under the North American Plate. Some forces are rapid, like the monumental Miocene flows of molten basalt that formed many headlands.

Roger and I begin to walk along the Oregon Coast Trail on a beach at the end of the road from Tillamook. Walking south, we step over cobbles and driftwood, then walk along dry sand as we head to Cape Meares' prominent headland. The trailhead switchbacks up steeply through a Sitka spruce forest. The Sitka spruce with their large scaly patches of bark stand tall, with younger western hemlock below growing on decayed logs. The trail bends again and again heading up the slope, taking us through gooey muddy patches; openings show us glimpses of the beach below. Up on top of the Cape Meares headland, we encounter a gigantic spruce—forty-eight feet around. Down below, breaking waves rhythmically pound offshore volcanic rocks.

Sitka spruce are present everywhere along our coastal hike. This tree grows in a temperate rainforest along a narrow coastal strip from Prince William Sound, Alaska, to Northern California, where the climate is mild. Sitka spruce tolerate the continual sea spray. Growing in the understory are hot-pink flowering salmonberry, shiny-leafed salal bushes, delicate deer fern, and the heart-shaped leaves of false lily of the valley.

Oregon's coastal beauty was saved from overdevelopment over fifty years ago when Oregon's beaches and coastline were made public property by Governor Tom McCall, a Republican, in the "Beach Bill" of 1967.[1] The Oregon Coast Trail, connecting all seventy-five coastal state parks, took forty more years to complete. This trail gives us access to nearly all beaches along the entire 367 miles of Oregon's coast.

I can only imagine what Oregon's coast would have been without such protection. I've been to Florida, where most beaches are privately owned and the shoreline is crowded with high-rise condos and hotels. Along Washington State's coastline, many beaches are privately owned. While visiting the San Juan Islands near Seattle, we encountered "Private Property!! Do Not Enter" signs all along the shoreline. Very unwelcoming. By the time Washington stopped selling its tidelands, about three-quarters of its shoreline had been sold off to private landowners. The coastline, like many natural entities, is a limited resource.

Only 175 years ago, most Americans believed their frontier was endless, containing unlimited bounty. Cutting the vast swaths of native forest was justified as helping "civilize" the land. But those utilitarian, careless approaches to the natural world began changing in 1863 with a book—*Man and Nature.* George Perkins Marsh, its author, witnessed how Vermont's once-continuous forests were cleared. Influenced by New England's transcendentalists, Ralph Waldo Emerson and Henry David Thoreau, Marsh wrote about Western man's destructiveness wherever he set foot. When nature is destroyed, it remains impoverished until restored. "Restoration would redeem the landscape, and as man became a co-worker, would redeem man as well," Marsh wrote.[2] He argued that if we failed to halt indiscriminate logging, civilization would collapse, just as the Roman Empire had fallen. Marsh described the corruption of government that colluded with private corporations and urged people to take personal responsibility. His book became an international bestseller. sparking the establishment of forest reserves, inspiring national parks, national forests, and national wildlife refuges. Teddy Roosevelt, an avid hunter and conservationist, used his position as president to protect over 230 million acres of public land. Although he had objectionable views toward people of color, he popularized taking trips to the wild outdoors.[3] He wrote that by visiting wilderness, Americans would not miss the disappearing physical frontier.

Human activity is causing a modern global extinction crisis. Reversing these losses will depend on us engaging in conservation and restoration activities. Environmental writers who, like Marsh, urged us to pay attention to the natural world still inspire us. We look to the insights from geology to appreciate the enormous passages of time that have shaped the landscape, and to ecology to appreciate how elements of each ecosystem work together.

CAPE MEARES TO BOB STRAUB STATE PARK

Descending from the Cape Meares headland trail, we reach Bayshore Drive and walk along it to round the headland and reach the beach again at the small town of Oceanside. Once there, we continue walking along on the firm wet sand. Our dogs, Nala and Lulu, gleefully run up ahead and return, running back and forth for miles down the beach. Homes up on the headland face the sea. Looking to the south, there is the next headland, Cape Lookout, a long peninsula leading out toward the ocean. Other people on the beach are bent over and searching for clear or white agates. Agates are abundant among the volcanic rocks here; they form inside their cavities. The sun sparkles on the breaking waves as we walk.

All around us, in three directions, is sand, the wide-open sky, and the sparkling sea. It's a glorious day. Reaching the end of the sandy stretch of beach, we round a promontory where Netarts Bay meets the ocean. Waves traveling into the inlet break inside the bay; the tide is turning. Opposite us lies a long sandy spit that protects the shallow bay. Seals are hauled out there in a brown and gray mass of bodies. The spit is remote, and with no people to disturb them, the seals bask contentedly. A line of brown pelicans soars overhead; three pelicans plunge-dive into the water, having spotted fish, and reemerge. Then they float on the surface a while. We pass vacation cottages built up above the coastline. The tide within the bay slowly rises with each successive set of waves.

We head to Lookout Point State Park to hike up the North Trail. The trail follows the beach, then zigzags up a series of switchbacks to a terrace above the beach. We continue ascending the bluff though a Douglas fir and Sitka spruce forest. The deep *quork* of a raven startles us as we enter a section of old-growth forest and hike up the steep trail. Through breaks in the trees we get glimpses of the sparkling surf rolling up the beach. We pass more enormous Sitka spruce growing along the bluff.

These giant old-growth Sitka spruce are four to eight feet in circumference. At the base of their trunks, they flare out by as much as ten feet. Wind damage to the trees' crowns has induced buttons of canker to grow, each one fuzzy with tight balls of licorice fern. Smaller trees—western hemlock, red alder, bigleaf maple—grow under the canopy. Hemlock grows atop stumps and fallen logs, on the remains of conifers from the past. Where sunlight strikes the forest floor, thick clumps of fern and vine maple, salal, hazelnut, and huckleberry grow profusely. Impressively large moss mats cover the ground, blanket the downed logs, and coat the limbs and trunks of deciduous trees. Shiny broom moss takes the sunniest spots, covering logs and the bases of deciduous trees with their yellow mats. A couple of switchbacks take us farther upward to an even more majestic stand of old-growth Sitka spruce. Impressively wide, they stand like silent old beings, guarding the forest.

"I THINK that I cannot preserve my health and spirits, unless I spend four hours a day at least...sauntering through the woods and over the hills and fields," wrote Henry David Thoreau.[4] Spiritual truths were revealed to him as he walked, about human freedom, self-expression, and self-realization. If only more of us went out, open to nature's revelations, humans could align themselves more towards the "good," Thoreau believed. He wrote that walking helps us remember our own innate wildness. On long walks, as you float along, he theorized, you become a part of where you are walking through. In his essay "Walking," he insisted that we saunter "through the fields and woods for the sake of harmony," not merely because we want exercise.[5] We are not fully alive if we remain in the city, he believed, because city life tames us. Since human institutions are corrupt, and nature is pure, he believed the divine could be revealed only through nature. He opposed a materialistic lifestyle and advocated for a life of radical simplicity, which we might call sustainability.

Thoreau was influenced by the relationship between Native people and the natural world. Before he wrote *Walden* and *The Maine Woods,* he spent time with the Penobscot Indians, as their lives were being overturned by commercial interests. He wrote about tribal cultures while at Walden Pond and while traveling to Minnesota. He read Henry Rowe Schoolcraft's *History of the Indian Tribes of the United States.* He listened to Native people while in the Maine woods and out on the Minnesota prairies. Two of these people, Joseph Attean, a Penobscot tribal member, and Joseph Polis, with whom he hiked in the Maine woods, are mentioned.

Thoreau wrote about loving to listen to Indian languages and hearing their histories and traditions around evening campfires. He wrote that his Indian companions saw the "Great Spirit" in everything, and demonstrated both respect and reverence for nature. These experiences influenced his own developing philosophy and helped him understand the spiritual values of wilderness, the unity of all life, and the need to live harmoniously with nature. Almost all of what he wrote about this was left in his "Indian Notebooks" when he died. He intended to publish them to correct the nineteenth-century Euro-American misconceptions about Native people.[6] A selection of these notebooks has been published.[7] Reportedly, Thoreau spoke two words before he died: "moose" and "Indian."

Although Thoreau wrote about wilderness, the landscape he encountered was not untouched in the sense he believed it to be. Native tribes had shaped the landscape for thousands of years before him. Native people used fire and flooding to manage their landscapes for their preferred foods, intentionally increasing the diversity and productivity of the land. Euro-Americans clung to a belief that Native Americans were primitive, environmentally benign, and incapable of shaping their habitat in such ways. The touch of Native Americans to the land was gentle compared to that of the timber barons of the nineteenth century. Timber companies cleared the forests around the Great Lakes by the late 1800s, without replanting them, then set their sights on the forests of the West Coast.[8] Until relatively recently Euro-Americans did not recognize how Native Americans possessed a sophisticated knowledge in managing the land.

CAPE LOOKOUT is a narrow, eight-hundred-foot-high, two-and-a-half-mile-long peninsula, jutting into the Pacific Ocean. Its forest is a puzzling quilt of different-aged stands of forest—young, mature, and old growth—which suggests that windstorms and other ecological disturbances have been common occurrences during the last two hundred years.

Reaching the top of the headland, we look out over the ocean. The approaching rows of rolling waves beckon us downward to a long stretch of sandy beach. The dogs eagerly lead us down the switchbacks of the South Trail. Maybe they smell the tangy sea air? Spruce and hemlock roots cover the trail. The rapidly descending, flutelike notes of a Swainson's thrush ring out along the way. As we reach the lower part of the cliff, sword fern is growing everywhere. The air is warmer now, protected from winds by the headland, and smells vaguely tropical. Our final descent takes us along shorter and shorter switchbacks through the trees and shrubs until we reach the beach. Directly north stands the tall, dark basalt cliffs of Lookout Point. Formed by a massive volcanic event, the basalt hardened from lava originating as a plume of magma from deep in the earth's mantle.

These towering basalt headlands mark where massive overland basalt flows poured into the sea fifteen million years ago. I try to imagine what these enormous outpourings of lava from a rent in the earth hundreds of miles to the east looked like. Somewhat like the outflow from Kilauea volcano, but a quarter of a million more cubic miles of lava. These Columbia River flood basalts, as they are called, spread throughout the Pacific Northwest, flowing down what would become the Columbia Gorge, heading toward the Pacific Ocean. The lava moved slowly, a few miles an hour, filling ancient river valleys, pushing the ancestral Columbia River northward to where it is today. Probably, it flowed through enormous insulating lava tubes that helped keep it moving. The lava covered parts of Oregon and Washington under a thickness of thousands of feet—the volume could have buried the entire United States beneath forty feet of lava.[9] There were as many as twenty of these flows, and several reached the Pacific Ocean to create headlands—called capes—like this one. When it reached the Pacific Ocean, the lava cooled and hardened into fractured pillow-shaped basalt. The enormous basalt lava floods buried the older rocks beneath them.

The Pacific Northwest as we know it did not exist 200 million years ago; the Pacific coastline ran through Idaho at that time. The oldest rocks in Oregon were once volcanic island arcs, occurring far out in the ocean, riding on crustal microplates. The most recent add-ons were microplates called the Siletzia terrane.[10] Oregon added itself to the North American Plate about 45 million years ago. In many places, accreted terranes are hidden under younger rock. We will see evidence of accreted terrane for ourselves where they are exposed in southern Oregon.

Impressive as Oregon's Columbia River basalt floods are, they are dwarfed by the older Deccan Traps flood basalt flows in India, an event that helped doom the dinosaurs to extinction. So much lava flowed, emitting so much sulfur dioxide and mercury, that world temperatures plunged by 36°F, acid rain fell over the earth, and the ocean acidified.[11] Even more cataclysmic were the series of basalt flows in Siberia, whose sulfur dioxide emissions were so intense that they led to the great Permian extinction 250 million years ago. Most shallow water marine animals, like trilobites, went extinct.

The waves are turquoise green in the sun as they lift to break. The beach curves gracefully south toward Sand Lake. Beyond are the tan bluffs of Cape Kiwanda. Farther down the coast is the rounded promontory covered in grassland called Cascade Head, a last refuge for the endangered Oregon silverspot butterfly.

South of Lookout Point is Bob Straub State Park, where we continue walking south along a broad, white sandy beach. Oregon's coastline is overlain with sandy beaches along two-thirds of its coast. Inland rocks erode into pieces from rain and frost action and are washed downstream. Eventually decomposing into sand, millions of tons of it wash downriver to the sea. Once the sand reaches the sea, longshore currents move it up or down the beaches.

DURING THE ICE AGES, Oregon's coastline was ten to forty miles farther out to sea. The rivers and streams extended out this same distance. When the ice melted, the mouths of these rivers were drowned. Nestucca Bay is one of the many bays and estuaries that was thus formed. The white sand beach we walk along gradually thins to a point as it meets the mouth of the bay. On the opposite side of the dunes from where we are on the beach is a wildlife refuge. Migratory waterfowl, beaver, mink, and river otter live in the marshes, and salmon and trout live in the bay.

Reaching the mouth of the estuary, we turn inland to walk along the edge of the estuary. It's high tide, and the small embayment at the mouth is alive with pelicans, gulls, surf scoters, white-winged scoters, and double-crested cormorants. A line of thirteen pelicans flies low over the water; one tucks in her wings and plunges into the water. The bird reemerges and sits on the surface and shakes its head, maybe swallowing a fish. More pelicans dive. When a pelican plunge-dives, her pouch expands with about two gallons of water and fish. When the pelican comes back to the surface, she lets the water flow out before swallowing the fish with a toss of her head. With several strong wing-beats, the birds alight again and move back out toward the ocean.

Scientists have drilled down into this wetland to extract soil cores and learn about its prehistory.[12] Some retrieved cores contained marine diatoms, brought inland by three past tsunamis. Out at sea lies the Juan de Fuca Plate, which stretches along the coastline of Oregon and Washington. The zone between it and the North American Plate is called the Cascadia Subduction Zone. When the Juan de Fuca Plate jerks forward under the North American Plate, the subduction zone snaps; massive earthquakes with long and severe shaking ensue. Earthquakes generated in this way—megathrust earthquakes—occur every three hundred to five hundred years. They are followed by massive tsunamis. The most recent massive tsunami washed seawater miles inland here. We know the date, January 1700, because those same tsunamis reached Japan, where meticulous records were kept.[13] How bad was the earthquake? A Tillamook tribal member described the earthquake as a violent "shaking of the mountains so that it was impossible to stand," according to a written account from that time. And the tsunami? A Klallum Indian described the rivers becoming salt, the valleys filling with water.[14] These tectonic plates will slip again, the fault will snap, causing tremendous shaking, tsunamis, landslides, and liquefaction of loose soils and collapse of any buildings built on them. After somberly considering this fate, we turn around to walk back from the mouth of the outlet and along the beach.

We head south in our car—past tiny Winema—to another stretch of beach. Walking three miles along the flat stretch of sand toward Neskowin, we watch as flocks of sandpipers forage in the wet sand behind receding waves. The waves approach the shore evenly, gleaming green-blue as they rise and break into the surf. How good the warm sun feels.

OTTER CREST

There is a minus tide along the shore this spring morning. Yesterday, at the precise moment when the sun and the moon were aligned on opposite sides of Earth, we saw the full moon. The ocean's response is slowed by friction, water's inherent inertia, how deep the ocean is, the position of the continents, and even by the earth's rotation. Now, at eight this morning, the tide is low.

RACHEL CARSON loved exploring the intertidal zone during minus tides while at her summer house in Maine. Being a biologist, she would descend to the intertidal to collect some animals to show to her nephew or to inspect via microscope. A conservationist, she would return the living animals back to the tide pools where she had found them. She was an avid researcher, and once she was onto an idea, she could ceaselessly pursue it. She immersed herself in learning something new, starting by bringing home armloads of library books. She went out into the field and even went to sea on research vessels to dive, wearing the heavy metal diving helmet of that time, to observe firsthand.

Her job writing brochures for the Fish and Wildlife Service helped her develop a flair for popularizing natural history. She had always wanted to become a writer; other nature writers had inspired her. Henry Beston was one of them. In *The Outermost House,* he explained why we should respect animals. "Gifted with the extension of the senses we have lost or never attained, living by voices we shall never hear . . . , they are not brethren, they are not underlings: they are other nations, caught with ourselves in the net of life and time."[15]

It's stirring to turn off a human-centered perception and see other animals not as things but as other nations, with whom we need to ally ourselves. Carson also read Henry Williamson, who brought the animal world alive for others in a series of books, including *Tarka the Otter.*

Rachel Carson's first book, *Under the Sea Wind,* was written to help readers understand the lives of marine animals. Her next book, *The Sea Around Us,* combined science and lyrical prose. "Fish, amphibian, and reptile, warm-blooded bird and mammal—each of us carries in our veins a salty stream . . . in almost the same proportions as in sea water," she wrote.[16] This book was a call to action about the threats to ocean life from human exploitation and pollution. She wrote about how ironic it was that the sea itself should be threatened by humans, a form of life that rose originally from the sea. She describes with prescience how human activity was destroying our own future and that of other living things.

Silent Spring, the book for which she is best remembered, warns about the dangers from indiscriminate use of the pesticide DDT. Brown pelicans had nearly disappeared due to pesticide accumulation in their bodies, causing eggshell thinning. Peregrine falcons, white pelicans, osprey, bald eagles, and even robins were all affected. After the book was published, the pesticide industry furiously attacked her. She was accused of being a Communist sympathizer, called a spinster with an affinity for cats, and her credibility was demeaned by calling her a "hysterical woman." She bravely countered her detractors with facts. Rachel maintained people had a right to know and should be kept safe from lethal poisons.[17]

In a curious coincidence of calamities at this time, the public was concerned about nuclear fallout and the tens of thousands of babies born with malformed limbs and other deformities because their unsuspecting mothers had used Thalidomide while pregnant. Prescribed as an innocent cure for morning sickness, the drug was sold to pregnant women without adequate testing. The babies born deformed were grim reminders about how we were poisoning ourselves. The book had an explosive impact. DDT was banned from use in the United States. Public outrage over being poisoned led to new environmental laws protecting us from dangerous pesticides.

Seven years after the publication of *Silent Spring*, in 1969, the Cuyahoga River in Ohio caught fire and a massive oil spill in California despoiled the beaches of Santa Barbara. Waking up to the reality of environmental problems, groups of people launched the modern environmental movement. Powerful new laws, the Clean Air Act, the Clean Water Act, and the Endangered Species Act, were passed in Washington, DC. Rachel Carson did not live to see this happen; she died of cancer two years after her book was published. Public activism, the work of determined individuals, and the powerful Clean Water Act have since helped restore the Cuyahoga River.

What would Rachel Carson think if she were alive today? Brown pelicans, osprey, and other birds have rebounded. But we have new problems. New pesticides are being used, like neonicotinoids, which are a thousand times more toxic than DDT. These kill pollinators as well as pests. Pollution from caffeine and drugs improperly flushed down the toilet or leaked from septic tanks pour out into the ocean.[18] Blood thinners, endocrine-disrupting chemicals, heart medications, hormones, and opioid painkillers are in seawater.[19] Caffeine in the water stresses filter-feeding organisms in the ocean. Prozac, a common antidepressant, in the water makes crabs more likely to be caught by their predators. The ocean has warmed from human-caused CO_2 pollution. Coral reefs around the world suffer from massive coral bleaching. Dead zones—large stretches of ocean that lack sufficient oxygen for life—keep appearing off the coast.

ON THIS LOW-TIDE MORNING, it is cold and overcast. Roger and I park at Otter Crest Beach and head down a set of steps to the sea. We watch for a few minutes as the tide pulls away, exposing the low-lying rocky intertidal platform. The sound of the surf and the smell of the sea pervade the morning. All around the exposed rocks lies a field of flattened algae. We walk carefully out toward the retreating surf, stepping over primitive blue-green algae, lobed and olive-green *Fucus,* then the rough-coated *Gigartina,* looking like a red washcloth. Black oystercatchers make loud, high-pitched calls as they fly by along the shoreline. We stop to gaze down into small rock canyons filled with giant green anemones, looking like flowers, each "petal" bearing stinging cells. Many purple sea urchins fill entire tide pools. Little fish called sculpins dart near the walking shells that house hermit crabs. Small bits of bright pink sitting in the crevices first appear to be bubblegum worms—but, looking closer, Roger sees that they are actually bits of pink plastic. Bloodred sea stars adhere along the tide pool walls. Finally, we reach the lowest exposed area where red iridescent *Iridaea,* bullwhip kelp *Nereocystis,* and brown strands of *Laminaria* lie prostrate and exposed. I lift some limp strands. Colorful sea slugs are hiding under the kelp strands. Called nudibranchs because they have unprotected gills lining their backs, they defend themselves by eating and appropriating the stinging cells of anemones, then use the stinging cells for defense and to catch their own prey. Here there are orange-splotched and lemon-yellow dorids, and an opalescent nudibranch whose feathery gills are entirely bright orange. Farther out, still being pounded by the surf, is the odd-looking palm-tree-shaped kelp *Postelsia.*

THE INTERTIDAL is a place of transformation for me. As a young adult, I left college before my senior year to head to California. I had been a student for sixteen of my twenty years. I wanted to experience life instead of living life through books. I wanted to see the redwoods.

Reaching a small Northern California town, I rented a cabin on a bluff overlooking the Pacific Ocean. The sound of sea lions barking and the waves breaking on the rocks below was a constant presence. I would descend to the beach to the intertidal zone, observing animals that looked like plants and plants that looked like animals. What were they called? One day, I found a book—*Between Pacific Tides,* by Ed Ricketts and Jack Calvin—in the bookmobile visiting our town. Ricketts amassed his knowledge through direct experience, standing thigh deep in the water of the intertidal zone. He also wrote well. Bringing the book home, I avidly read about the unique lives of the intertidal animals and plants. My curiosity was reawakened, and I returned to the university to study biology.

On this morning, small groups of high school students moved among the slippery rocks toward the sea's edge. Some of them crouch around a tide pool, laughing as they point out animals to each other. There are multiple purple sea urchins in nearly every crevice on the marine terrace. We approach a small group talking about sea otters with their teacher. They know how sea otters could keep this urchin populations in check and allow offshore kelp forests to prosper.

Thousands of kelp plants stand like trees, anchored to the sea floor a hundred feet below, swaying in the current. Considered within a well-known ecosystem process called a trophic cascade, otters are predators that help limit the kelp-eating sea urchins and so keep kelp abundant. Without any predators, sea urchins cover-graze kelp and can become so destructive that they strip all the kelp from an area. Sea otters eat about a quarter of their weight each day in food, and their preference for sea urchins helps the kelp forests prosper and expand. When otters are present, fish depending on the kelp forest are more abundant. Sea otters could help buffer some of the negative effects of climate change. Kelp helps store carbon, in some estimates by as much as 1,867 pounds of carbon per acre of kelp. Restoring sea otters would also help reduce the CO_2 in seawater that leads to ocean acidification.[20]

We are near the place where the last sea otter in Oregon was hunted. Hundreds of thousands of southern and northern sea otters once lived off the Pacific Coast, along the narrow band of shoreline from Northern California through Alaska. Otters have the densest fur of any animal, with more than a million hairs per square inch. Sea otters off the Pacific Northwest coast were slaughtered for their dense fur until the late nineteenth century. Sea otter hunters sold the skinned pelts to the Hudson Bay Company. Captain Robert Gray, famous for changing the Wimral River's name to the Columbia River, had been in the area to buy up sea otter pelts when he made his geographical "discovery" of the Columbia. The Chinook Indians knew about the river long before Gray. The greed and plunder of the commercial sea otter skin trade nearly caused their extinction.

By the early twentieth century only one group of about thirty southern sea otters was left. Their presence off the coast of Big Sur, California, was kept secret until the otters were legally protected. From this remnant of survivors, otter populations have recovered and moved into some of their former grounds in Northern California. Alaskan sea otters have expanded down the coast to Washington State. Fishermen who harvest sea urchins, clams, and crabs are not happy about the revival of sea otters; some have shot and killed otters who were taking seafood the fishermen felt they were entitled to.

Happily, environmental organizations and Native American tribal groups have formed an alliance named Elakha—"otter" in the Chinook language—to reintroduce sea otters to Oregon.[21] Scientists have mapped the kind of offshore habitats sea otters in California prefer to predict where on the Oregon coast they might eventually settle. Depoe Bay, near here, and southern Oregon sites near Port Orford and Gold Beach are all optimal habitats for future sea otter reintroductions.

We head a short way up the coast to Depoe Bay, where we pull over and scan the horizon for the heart-shaped blow of whales. Gray whales by the thousands pass this way along their long migration between Baja and Alaska. Several hundred whales spend their summer here off the Oregon and Washington coast. They are affectionately called the Pacific Coast Feeding Group.

Gray whales tend to stay where food is rich and water is deep, even close to shore. They slurp up mud and filter it for tiny shrimplike amphipods by forcing the mud through their baleen. The baleen traps the food, and the whale uses its enormous tongue to swipe the shrimp down its throat. By mud foraging, whales also mix nutrients into the water; the feeding trails in the sea-floor they create benefit other sea animals. Gray whales also chase down swarms of fish to eat, or might scoop up swarms of tiny mysid shrimp. Eating anytime, night or day, sometimes they use the noise made by their prey to locate them. Snapping shrimp, for example, make loud snapping sounds that create air bubbles. The bubbles help the shrimp stun their own tiny prey. Whales can zoom in on the snapping sound, says Oregon State University marine acoustical researcher Joe Haxel, to gobble up the shrimp.[22]

WHALING WAS BIG BUSINESS, and very profitable. During the 1860s, a gallon of whale oil was worth $1.92, and a barrel of whale oil was $1,500.[23] Many whaling grounds were abandoned by the end of the 1800s because the whales were nearly extinct. The American Pacific Whaling Company, later becoming the Consolidated Whaling Corporation, concentrated on hunting gray whales on the west coast of the United States. Whaling was risky, so corporations spread the risk by having investments by shareholders. This model of doing business was later emulated by modern corporations.[24] Charles Scammon discovered the gray whale calving grounds in Baja, Mexico, in 1860; he slaughtered the whales there. Norwegian whaling companies had a whale oil cartel and operated some of the first factory ships in the Pacific; when a whale was killed, it was hauled aboard, flensed, and boiled down for oil.

Killing gray whales continued until 1938, when only 250 animals were left. Whale oil was replaced with fossil fuel, and an international ban on whaling was passed in 1982. The population has since recovered, with about 25,000 gray whales now—a conservation success story. But all is not well. As ocean temperatures in the Bering Sea become warmer, the water becomes less productive. Increasing numbers of gray whales have washed up, emaciated, onto the beach. No one knows the exact reason; perhaps their favorite food, amphipods, is becoming less available due to climate change.

What would Rachel Carson do? Write a new book? Testify in Washington, DC, for strong climate protection laws? Would she encourage more children to develop a sense of wonder about the world?

HECETA HEAD

About halfway down the Oregon coast we pass through the small coastal towns of Waldport and Yachats. By the side of the road is a small handmade sign, suggesting "Choose Kindness!" Somehow, that sign moves me to tears.

A REMNANT GROVE of old-growth Sitka spruce stands at Heceta Head. Beginning at the trailhead, we head toward it on a carpet of beaked moss and lichen. This forest is abundant in moss and lichen, having over fifty-seven different species counted in just a one-acre plot. There is tiny false pixie cup lichen on the ground, and curtains of feathery cat-tail moss hanging from branches, along with dangling strands of old man's beard lichen. Bright green bits of *Lobaria,* or lungwort, have fallen from the treetops and lie on the ground. People used to believe this lichen could heal pneumonia because it is shaped somewhat like a lung. Believers in the doctrine of signatures thought that God was showing them useful medicines by shaping plants like the organ they would help heal. This lichen is only present in old-growth forests when the air is clean; it cannot tolerate polluted air.

We pass dense stands of young Sitka spruce as the trail climbs toward the headlands. Where is the old-growth spruce? I wonder if we are in the right place. But we continue and finally reach the first of many very large, old trees. This forest is old, and tree trunks have grown twisted and extraordinarily wide in circumference. The layers of the forest blend into one another. Up high are the large spruce branches; beneath them, growing in sun gaps, are the young spruce, hemlock, alder, and maple. Below those grow native rhododendrons, now in bloom with bright pink blossoms. Thickets of salmonberry, vine maple, tall salal, and evergreen huckleberry grow nearby. On the forest floor are scattered piles of dead wood and abundant moss and lichen. Bunches of sword fern, bracken fern and deer fern, and *Angelica* grow up from the forest floor. This rainforest is so biologically productive that scientists say its biomass—total weight of all living things—is the largest of any ecosystem. There is plentiful rain—two hundred inches in a year—and mild temperatures.

Tall and huge in diameter, each spruce tree sprouts enormous epicormic branches. These branches grow from a dormant bud after the trunk is exposed to increased light or is disturbed by a ground fire, like what happens to redwoods. Unlike with redwoods, these epicormic spruce branches begin low to the ground and curve upward, forming giant candelabra shapes. The branches are as thick as the trunks of the younger trees we had passed earlier. Many of these old trees support giant clusters of sword fern and licorice fern in nooks created between the main trunk and the huge branches. There are dozens of these old trees all around us.

JOHN MUIR, an advocate for wild forests like this one, is famous for having taken a wild ride atop a giant fir during a Sierra storm. Climbing the tree, intending to experience the divine power of the storm, he clung on as its crown circled and gyrated in the high wind. From his perch, he looked out over the valley, seeing and listening to the power of the wind as it stirred the trees: "running in ripples and...swelling undulations across the valleys from ridge to ridge.... The sounds of the storm...the profound bass of the naked branches and boles booming like waterfalls."[25]

John Muir fought to protect wild areas, but those areas were enjoyed by mostly white well-off tourists. Muir and those conservationists who helped preserve the redwoods considered their own European backgrounds to be superior to other cultures. During John Muir's time, Native tribes had little legal standing. Miwuk tribal members, who lived in Yosemite Valley, had to negotiate to be able to stay on the land comprising Yosemite National Park; many other tribes were evicted when a national park was created.[26]

WE TURN ONTO A TRAIL heading to the beach. As we descend, the branches from moss-draped maple, salal, and stunted spruce trees on both sides meet overhead, forming a tunnel. The trail drops down to the secluded beach. A stretch of white sand spreads to the surf; to the north are offshore rocks where animals are protected in Oregon's Islands National Wildlife Refuge. Both seals and sea lions are hauled out on the distant offshore rocks; the sea lions are barking loudly. Waves strike the rocks from behind, splashing up into the air, then streaming over the rocks like a champagne fountain at a wedding.

We look up where the big trees grow like a spiky crown atop the headland. On the beach, a young woman wearing an oversized backpack approaches. She leans forward with the effort of walking in the sand. We all stop when she nears. "Hiking the Oregon Coast Trail?" I ask. She responds wearily, "Yes." She is hiking the entire trail solo. Today she has gone about ten miles, mostly in the sand. It's the twentieth day of her hike and she is just halfway along the route. Her voice is soft, and she seems withdrawn. We all sit together for a few moments, passing around a bag of trail mix.

EXPERIENCES IN THE WILDERNESS can help a young person build their self-confidence. My friend Angie has guided many different groups of young people through a wilderness challenge program. They learn a set of basic outdoor living skills before heading to the wilderness. Once there, they live in often adverse conditions using only what they carry on their backs over weeks at a time. Angie has described the transformations that she witnessed. One group was hiking through a July snowstorm. They had to work together, trusting that they could handle this and trusting one another. Gradually, they got through the adverse conditions and reported later how they had felt more alive and whole throughout this experience. As they grew increasingly animated, they began to perceive nature as being just as alive and whole.

Wilderness experiences can have lasting impacts. Youth who had participated in programs like Angie's were interviewed after one, five, and ten years had passed. The researchers systematically sampled programs of different lengths and different age groups. The participants continued to value wilderness even after ten years had passed. They still remembered details about how their own life perspectives and self-awareness shifted during their wilderness experiences.[27] Wilderness experiences are significant for a young person, possibly because of an expanded sense of freedom. Researcher Stephen Kellert says people's normal habits and sense of who they are changes with time spent in wilderness. Kellert and biologist and author E. O Wilson think being in wild nature has positive affects due to "biophilia." What we observe in nature reflects our own inner, deep mental patterns.[28] They reason that since humans were shaped by long expanses of evolutionary time spend in wild nature, time spent in wild areas reawakens positive feelings about nature.

WE STOP IN FLORENCE at the long jetty extending beyond the mouth of the Siuslaw River. If we had been here at the end of the last ice age, an ancient bay, called Heceta estuary, would have appeared before us. The sea level was then as much as four hundred feet lower. This ancient body of water would have been larger than San Diego Bay, with a protective arm bending around it another thirty-five miles to the west. Huge, rushing rivers would have run down to the bay. The estuary would have been full of salmon, crabs, and other shellfish, juvenile marine fish, marine mammals, and birds. It would have been a perfect place for Native Americans to settle.

Thirteen thousand years ago, when the Heceta estuary was exposed, it was part of a coastal corridor—the kelp highway—along which early Native Americans traveled down the west coast using watercraft. This waterway had opened a coastal route from Alaska down the Pacific Coast about fifteen thousand years ago as glaciers began melting. The ancestors of today's Native Americans traveled the kelp highway, staying on to live in choice locations.[29] Their movement down the coast would have been helped by how familiar the people were with marine food resources. The shellfish, fish, sea mammals, seabirds, and types of seaweed were similar all along Pacific Rim coastlines throughout North and South America.[30] Since the end of the ice ages, rising sea levels have submerged entire coastal regions. Almost all of the continental shelf coastlines where these people might have lived are underwater. Much of Oregon's former shoreline is now under four hundred feet of water.

The underwater ancient Heceta estuary likely contains sites where early humans once lived. Archaeologists search below sea level for archaeological remains using underwater sonar to provide a level of three-dimensional exactitude down to the nearest meter. Their hunches are supported by what archaeologists found near the town of Bandon: stone flakes, blades, and charcoal from fires in a deposit thirteen thousand years old. Humans had lived here along an inland coastal stream and dunes.[31]

SAMUEL H. BOARDMAN STATE PARK

We pass Gold Beach along the south coast and stop at a beach near Cape Sebastian. Here we are looking for chunks of ancient deep-sea floor that has been shoved up onto land. The oldest rocks that would become Oregon were once volcanic islands formed somewhere in the Pacific basin. Riding on microplates far to the west or to the south, they were carried along as these plates shifted west and north. As the floor of the Pacific Ocean was subducted under North America, these terranes were too big or too thick to slide beneath. They were scraped up and "welded" to the western edge of North America as the oceanic plate slid under the North American Plate. These terranes accreted onto Oregon's landmass accompanied by chunks of former ocean floor. Ancient sea-floor finds itself high and dry on the beach.

These assemblages of rocks are called accreted terranes.[32] Each terrane is made up of unique sequences of bedrock, very different from the rock that now surrounds it. Each terrane is separated by faults. Some terranes originated in tropical settings; made of limestone, it contains tropical coral fossils. We are looking for blue-green serpentine and a pillow basalt from ocean crust, and basalt from ancient volcanic seamounts. There might also be crushed and broken rock called mélange formed by the grinding together of the tectonic plates. Complex stresses around subduction zones have deformed each terrane; with heat and pressure, the rocks metamorphize. They arrived here, one after the other, gradually converging to construct the Pacific Northwest.

When we examine the sea stacks sitting on the beach, there is an exposure of peridotite, metamorphosed into serpentine. Peridotite is the dominant rock of the upper part of the earth's mantle, brought up from depths of 18 to 125 or more miles. There are also thick sheets of basalt, old hard sandstone and shale, all highly sheared. This area was once part of an island arc, like New Zealand but on a smaller scale, that arrived here from twelve hundred kilometers away to the south. Fossils within the sandstone and shale show the rocks to be 150 million years old.[33] They were formed during the Jurassic age, when the first bird—Archaeopteryx—was present. This famous extinct bird, about the size of a raven, had both feathers and reptilian jaws with teeth, along with three-fingered claws on its feet.

CONTINUING SOUTH to Cape Ferrelo, we reach Samuel H. Boardman State Park to visit a coastal prairie. This one grows thick with native perennial grasses: reed grass, oat grass, fescue, blue grass, tufted hairgrass, and needle grass. Some of these plants may have existed here for a hundred years or more. Perennial grasses remain green late into the year by absorbing summer fog, similar to how redwood trees utilize fog. These grasses extend their roots deep into the soil to access water. Tall bracken fern, wild strawberry, Douglas iris, blue-eyed grass no longer in bloom, frosted paintbrush, bright yellow sneezeweed, white clusters of yarrow, pearly everlasting, and purple lupine all grow here. Standing on the coastal bluff, we are surrounded by the calm Pacific Ocean in three directions. We head down among a thick cover of salal and evergreen huckleberry, coyote bush, forests of Sitka spruce, and a carpet of false lily of the valley.

Coastal prairie patches like this one were once maintained by Native American burning. Over a hundred and fifty years ago, Dr. Lorenzo Hubbard stood nearby to watch Native Americans burn a prairie not far away at the mouth of the Rogue River. He wrote that a local tribesman told him this practice of burning was performed to invite salmon to enter the mouth of the river.[34] Native people actively manipulated fire to maintain the plants they needed for food and medicine. Tribes still practice this management technique. In late summer, they carried out low-intensity burning of prairies when the meadows were dry. Burning encouraged the growth of camas, an important food for the people, by helping eliminate competing vegetation. Active participation in managing the land was bounded by a code of conduct, sometimes called the "Honorable Harvest." Robin Wall Kimmerer, a botanist and member of the Potawatomi tribe, explains how Native tribes benefited from their long-term observations of the land and its inhabitants, creating patterns of ritualized stewardship over time.[35]

Beginning in the mid-1800s, white settlers introduced agriculture, grazing, and alien plants. After "Indian" burning was forbidden in Oregon, trees began to cover the prairies and prairie plant and animal diversity declined. Now, most western coastal prairies are covered in trees and shrubs. Comparing sets of old photos of coastal prairies with modern ones taken from the same point, I see for myself how, without the benefit of fire, coastal grasslands become covered with trees and shrubs. Low-intensity burning used on some coastal prairies has helped generate biodiversity by multiplying the number of different patches. Each patch has different sets of plant communities, creating higher biodiversity and resilience to change.

Land needs to have someone living on it, according to Kimmerer. She recommends living in reciprocity with the land in a nondestructive relationship. Take only what you need, leave some for other beings, share it and give something back, she explains in her book *Braiding Sweetgrass*.[36] Kimmerer recommends seeing other living things as relatives, having as much animacy as people do. People are not the only beings who deserve respect. In olden times, the elders said the trees would talk with one another. Kimmerer reminds us that those elders were right; scientists now understand how trees communicate through pheromones and through their mycorrhizae.

We have been walking all day and decide to stay on the bluff as the sun sets. I have been watching sunsets for several years, hoping to see the rare green flash. As the sun sinks behind low clouds, it seems unlikely we will see the phenomenon tonight. But suddenly, the sun emerges into a clear opening of sky just at the horizon. We watch the orange disk slip away, and at the last moment there is the momentary gleam of brilliant emerald green, like a jewel. Then the light is extinguished.

CHAPTER 6

Rivers of the West

THE COLUMBIA RIVER, flowing through twelve hundred miles of Canada, Washington, and Oregon, is emblematic of our nation's trajectory. Once it was a rich watercourse supporting what were probably the largest salmon runs on earth. The river fed people and many aquatic and terrestrial species over ten thousand years. But in just a hundred and fifty years, salmon runs and the people who depended on the salmon were severely diminished. The river's power has been reduced to a chain of flaccid lakes lying behind sixty dams. Yet there is hope for a turnaround.

The river originates with trickling meltwater in the Northern Rockies of British Columbia. The outflow of the Athabasca Glacier runs over and under the glacier, forming streams of outwash, converging into a river. The ice is a relic from the Pleistocene, when a mantle of ice blanketed Canada and the northern United States in the west. The young river first runs northward and then turns abruptly south, joined by tributaries along the way, eventually crossing the Canadian-US border. The river meets the Snake River—the Columbia's largest tributary—in southeastern Washington. It then runs westward, continually joined by tributaries, passing The Dalles of the Columbia River Gorge. The river is joined by the Willamette River and turns slightly north again, flowing past the Ridgefield National Wildlife Refuge, eventually discharging into the Pacific Ocean. Its watershed is immense—the river and tributaries flow through an area the size of France.

We stand where the Columbia empties into the Pacific, at the mouth of this massive river. From the beach, we look across the river's lower estuary and watch as ocean waves crest and run up to mix with the river. During violent storms, ocean waves forty feet high can run up into the estuary. Today, a sturdy bright yellow tugboat marked "Pilot" heads out of the bay toward a huge freighter that lies in wait beyond the sandbar to help it navigate the rough and dangerous entry. Several hundred vessels have sunk and hundreds of people have died trying to cross the sandbars and treacherous currents here.

The Chinookan people had navigated along this river for thousands of years, but the river's entrance was obscured to European explorers by fog, currents, and shifting sands. Finding the mouth of the Great River of the West stumped explorers from Spain and England until American Robert Gray sailed up the river in his ship, *The Columbia,* while buying up sea otter pelts in 1792.

We walk along a sandy peninsula that juts into the Columbia estuary. Flicks on the water's surface indicate salmon are returning, entering the river mouth after spending several years out at sea. Dozens of fishermen stand thigh-deep in the water of the estuary, holding their poles. I ask one young man what is running. "Coho," he grins. Another fisherman a few dozen yards away shows me the small, squirming silvery salmon he has just caught. These salmon might have traveled as far away as the Gulf of Alaska to their feeding grounds. Now they are heading upstream toward their high-elevation, cold-water spawning grounds. On the reverse journey next year, young salmon who manage to make the downstream trip this far enter the ocean between May and August.

This river once hosted giant fish. A 1925 photo taken in Astoria, near the river's mouth, showed a fisherman posing with his eighty-five-pound salmon catch. The fish is as large as the fisherman. These giant fish disappeared after the Grand Coulee Dam, six hundred miles upstream, was completed in 1939.[1] The dam closed off half of the fish spawning habitat in the Columbia River Basin.

The Columbia River once sustained ten to twenty-five million fish each year. Salmon are fundamental to the identity of the Indigenous people who continue to live along the river.[2] Native people continue to fish there. Historically, the river people's cultural traditions constrained their salmon harvest. Northwest Coast Indigenous population density was high, and although the tribes caught millions of pounds of salmon each year, their fishing techniques were relatively sustainable. Researchers Chad Meengs and Robert Lackey examined the remains of past salmon harvests over the last 7,500 years and describe fish populations as being remarkably stable over that time.[3]

Historically, the tribes' belief systems and traditions prevented overharvesting. Rituals—like first-salmon ceremonies—are still held when the first spring salmon return upstream. There was no fishing until the ceremonies were completed. Indigenous stories tell of the origins of salmon, immortal beings who gave their bodies to the people so the people could live. There were customs about who could fish and where they could fish, which guaranteed the survival of enough returning fish to spawn. On the Klamath River, the tribes held rituals involving the construction and setting of fish weirs, which direct fish into a trap, during summer peak runs. These rituals and practices helped renew salmon abundance and contribute to a sustainable and resilient salmon fishery on the Columbia River and on the Klamath River up until the 1850s.

Diseases decimated the Native people, and they were forced off of their traditional lands onto reservations when the white settlers began to exploit the salmon. After only a century and a half, Euro-Americans nearly wiped out most salmon runs. New canning methods to preserve the fish developed; more than thirty fish-canning factories lined the river by the 1880s. Hundreds of thousands of cases of packed salmon were made each year. In 1927, the Bonneville Dam was built near Portland, and Grand Coulee Dam was built between 1933 and 1942 near Wenatchee, Washington. Now, over 75 percent of salmon returning to Pacific Northwest rivers, including the Columbia, were raised in hatcheries.[4] Nearly all runs of wild salmon in the Pacific Northwest are now endangered.

DAMS

Dams are perhaps the most enduring change made to American rivers. Dams were once such potent symbols of our technical might that more than seventy-five thousand were installed in U.S. rivers. As American environmental awareness grew, so did opposition to dam building. When two large dams were proposed for the Colorado River in 1953, right inside the Grand Canyon, David Brower and the Sierra Club mounted a huge national campaign against them and defeated the projects. Public opinion against dams has been growing since.

Dams block fish passage. About a quarter of all upstream migrating adult salmon don't make it beyond the first dam on the Columbia Bonneville Dam. Although this dam has fish ladders, if salmon manage to get up past the dam, they get caught in the stagnant water behind the dam. Many young salmon trying to get downstream to the ocean are killed as they reach the impounded water behind the dam. They are either gobbled up by predatory fish living in the lake behind the dam or are torn up by the turbines of the dam. Very few young salmon survive this gauntlet to make it out to the ocean.

All dams have a limited life span because the river moves sediment as well as water downstream. Silt, sand, and gravel are stopped by the dam and accumulate behind it. In a freely flowing river, the sediment as well as logs and root wads are carried along downstream to the ocean where the sediment becomes beach sand. Hundreds of smaller dams have filled with sediment and are being decommissioned. Dams also change the quality of the water for the worse. The lake of slow-moving or still reservoir water behind the dam heats up, leading to summer algal blooms and decreased oxygen levels for any organisms in the reservoir, including fish.

Removing a dam was once considered radical, but since 2012 more than 1,720 dams have been dismantled to restore river systems in the United States. Once a dam is removed, the river system and its spawning salmon quickly recover. The world's largest dam-removal project occurred on the Elwha River, whose headwaters are in the mountains of Olympic National Park. Two dams had been blocking fish passage for ten different runs of salmon and steelhead along the river's forty-five-mile course. While the dams stood, people would watch adult salmon repeatedly bash their heads against the concrete wall of the dam, trying to get upstream.

Native tribes and environmental groups lobbied for years to get the Elwha dams removed. It was the cost of retrofitting the two old dams to modern standards that convinced opponents of dam removal to approve its decommissioning. Both dams were demolished beginning in 2011. In the first season after the dams were removed, the salmon returned. Thousands of Chinook salmon were seen spawning in the river. Over the ensuing five years, the Elwha River flushed out most of the tens of millions of cubic yards of sediment that had sat behind the dams and restored itself.[5]

On a visit to Olympic National Park, we headed up the Elwha to see the site where the lower dam had been removed. We watched a video of this dam being demolished over nine months. First a gap was opened into one side of the dam, and as water from the reservoir began pouring out, tons of earth behind the dam structure began to move. The river was redirected from one side to the other several times as the reservoir emptied. Finally, the rest of the dam was quickly demolished with dynamite. After more heavy equipment recontoured the rock and earth, it was revegetated. We walk along a decommissioned road, leading through a forest to the site of the former lower dam. It's a scramble to get down a steep embankment and stand above the cliff where the dam once blocked the river. There are vertical brown stains along the cliff face, marking where the dam wall had stood. Below us, clear, cold water flowed in a series of rapids.

The Glines Canyon Dam had been eight miles upstream; its two-hundred-foot walls were taken down over eleven days. The video footage shows a mechanical excavator moving along the dam rim, notching out cement, as the reservoir discharges its water through the notches in cascades. Dynamite blasted away the other structures. Standing on one of the remaining pieces of dam abutment, you can look down into the canyon, see where the spillway had been, and watch the unimpeded water cascade over the rock wall.

We travel downriver to the mouth of the restored Elwha. There had been no estuary at the mouth of the river while it was dammed; the coastline had extended flatly along a stretch of rock. After the dam removal, tons of rock and sediments flushed downriver and came to rest at the river mouth, building up a delta stretching out into the Strait of Juan de Fuca. The river and its side channels become sinuous, making an elegant curve. We walk along the many paths through willow jungles as far as we can go, peeking out at water running in a small channel. Here, young salmon rest and feed before entering the Pacific Ocean for their adult lives. Coho salmon, chum salmon, bull trout, and steelhead have all increased in number. We walk along the beach, where the main channel makes one last curve before entering the Strait of Juan de Fuca. A colony of gulls sits and flies over the bay. Otters dive in the river, and beavers have returned. Elk use the river during winter, and eagles fly in to consume the abundant eulachon.

TRIBAL MEMBERS work to help restore rivers and riparian areas bordering the rivers, along with scientists, management agencies, and environmental organizations. They endeavor to bring salmon back and reinvigorate their traditional ways of being in the world.[6] Tribes worked to help remove the dam on the White Salmon River, a tributary of the Columbia, in southern Washington. Afterward, returning adult chinook salmon quickly swam upstream to spawn in hundreds of redds.[7] The river was also recolonized by steelhead in tributary streams.

On the Penobscot River in Maine, two dams were removed and fish passage was installed on other dams, opening a thousand miles of habitat to migratory fish. After additional dams in the watershed were modified, the total amount of hydropower production remained the same. New technologies are being used, such as turbines that allow better fish passage and methods to increase the dissolved oxygen that fish need.

Older dams can be retrofitted to generate more electricity and have less environmental impact. In one example, closed-loop pumped hydro energy storage can be used to store power for a longer time than batteries. Pumped storage involves connecting two reservoirs of water, one at a higher altitude than the other, that are not continuously connected to the river system. When there's surplus electricity on the grid, these facilities use that power to pump water from the lower reservoir to the higher one. When electricity is needed, such as during lulls in wind or solar power, the water flows back downhill, spinning a turbine to generate electricity.[8]

THE KLAMATH RIVER

From its headwaters in the Cascade Mountains of southern Oregon, the Klamath River flows down through the Klamath Mountains to the Pacific Ocean in California. We stand on the north bank of the river mouth, high above the Klamath, in awe of the river's graceful curving form. Lying across the river mouth is a broad gray sandy spit created by waves. The water level is low, even though it is November; there is no plume of fresh water visible as the river enters the sea. Over the years, the position of the mouth changes, opening at the near shore, in the middle of the sandy berm, and now over at the far end of the estuary. During times of low river flow, July through August, the mouth may temporarily close off to the ocean.

The mouth of the river feels very much like a person here. The Yurok Tribe recently ascribed personhood to the Klamath River. Tribal attorney Amy Cordalis says this gives the river the right "to flourish and to naturally evolve and a right to a stable climate free from human caused climate change impacts." She explains that anytime the river is harmed, for example, by a toxic pollutant in the water, the tribe could bring legal action to protect the river.[9]

The Klamath once teemed with salmon; it was the third-largest salmon-producing river on the West Coast. The Yurok caught salmon, lamprey, and sturgeon in the river. But after settlers moved into the watershed, mining, logging, and water diversion for irrigation took a toll. Present-day runs of Coho and Chinook salmon are a fraction of what they once were. The fall chinook run was so small in 2017 that the Yurok Tribe cancelled the fishing season for the first time ever. In order to continue their tradition of a first-salmon festival, someone had to go and buy a salmon at the grocery store.

The six dams built upstream between 1903 and 1962 have the greatest impact on the river. The lowest one, the Iron Gate Dam, is a huge earthen mound sitting in the river. No provision for fish ladders was made, effectively cutting off the entire three hundred miles of upper basin habitat for fish spawning and rearing. The impounded water in each dam heats up, raising water temperatures and contributing to toxic algal pollution. A significant amount of water from the system is removed each year for crop irrigation in the upper basin. Much of the irrigated farmland is leased from national wildlife refuges in the upper parts of the basin. In drought years, a large proportion of water is often diverted for farming and rearing cattle, leaving little water for salmon. This has resulted in massive fish kills.

Brook Thompson, a member of the Yurok Tribe, was seven years old when one of the worst fish kills occurred in 2002. "I remember seeing thousands of dead salmon washed up on the shore," she told me. There had been a drought, but too much water was diverted to the farmers upstream, too little left in the river for fish. Crowded conditions and warm water temperatures caused a parasite called Ich—short for *Ichthyophthirius multifillis*—and a bacterial pathogen—*Flavobacterium columnare*—to proliferate. Over thirty-four thousand salmon were killed. What happens upriver affects those living downriver.

Brook described her family's long-standing relationship with salmon. "Salmon are family members. Our family have been taking care of salmon for generations." Her family had lived on their land, now a reservation along the Klamath River, for thousands of years. The Yurok were never removed to a far-off reservation as many other tribes had been. They have remained in place, although on only a fraction of their original lands. "The river is our highway and our grocery store," she continued. Brook explained how important salmon are to the diet of the Yurok. "It's literally killing our people when they become disconnected from their traditional culture and don't eat salmon." Without access to their traditional foods and because of the loss of their cultural connections with the river, diabetes, alcoholism, and a high suicide rate plague the tribe. Many tribal people live in poverty, and the nearest grocery store is as far as two hours away. "Our tribe has no hesitation to protect salmon. If corporations have a legal identity and are given a voice, so should rivers. We want to take the dams down," she continued. "We want a bigger role in decisions about using the natural world and put to use our traditional ecological knowledge, from thousands of years of observation, data collection, and understanding."

The Klamath dams trap nutrient-rich waters in shallow reservoirs. Stagnant water behind the dams causes an increase in summer water temperatures. Massive toxic blue-green algae blooms in the reservoir threaten wildlife and human health. The algae blooms deplete oxygen, further degrading water quality and affecting stream health below the dams. The dams also trap sediment, everything from clean gravel to silt, keeping the material from moving downstream. The mouth of the Klamath erodes because not enough sediments reach it.

When the Iron Gate Dam was up for relicensing in 2006, the law required that the different stakeholder groups be brought together to agree to the terms. The Klamath basin is a large area, and tribes, fishermen, environmental organizations, farmers, ranchers, and federal agencies that manage land and wildlife in the basin were all involved. Dam operators were required to put fish passages into the dam, and tribal groups began to argue that the dams should be removed. Despite decades of conflict between these groups, recent events had brought people together. The fish kill of 2002 and the overall drop in salmon numbers had hurt the tribes and fishermen. Upstream, farmers needing water for irrigation became vulnerable to low-water years when they might get cut off.

These groups didn't find it easy to compromise at first. Farmers who negotiated with environmental groups received anonymous threats, and tribal members who participated in negotiations were treated coldly in farm towns. But as groups worked together over a number of years, friendships and camaraderie developed. The coalition eventually developed a river restoration plan called the Klamath Hydroelectric Settlement Agreement—KHSA.[10] When the utility company PacifiCorp recently stepped in with additional funds, the largest dam removal and river restoration project in the world moved forward. Four dams will be removed and the watershed restored for fish. The first of four dams was removed in 2023.

No one knows for sure what will happen when the Klamath dams are taken down. Although the Elwha Dam removal serves as a precedent, the Klamath River is fifty times bigger than the Elwha. Still, people expect that when the reservoirs and dams are removed, the river will restore itself and water quality will quickly improve. The fish allocations promised by treaty to Native tribes can be fulfilled. Restoration along the river will improve the lives, health, and economic well-being of those dependent on the river. Removing the dams will restore hundreds of miles of salmon habitat, allowing salmon, steelhead, and lamprey access to over four hundred stream miles of historic spawning habitat upstream of the dams.

REMOVING DAMS on the Columbia River and its main tributary, the Snake River, is even more controversial. According to author Steven Hawley, the Snake River was once the most productive salmon watershed in the Columbia River Basin.[11] Now, the number of fish throughout the river system is about a hundredth of the original population. There are four dams on the lower Snake River, which provide only a fraction of electric power. According to the NW Energy Coalition, some years they provide only 1 percent of the electric power to the grid.[12] The dams mainly allow barge traffic to move wheat along the river. Some of these dams have fish ladders, yet many salmon die when trying to migrate upstream as adults or downstream as juveniles. Conservationists and fishermen have been pitted against wheat farmers and barge operators. Conservationists point out the alternatives to barging; the grain can easily be shipped by rail. The small amount of hydropower those dams generate is outmatched by other sources that are less damaging, cost less, and are more energy efficient. Wind turbines now produce more energy than is supplied by the lower Snake River dams. If these dams were breached, endangered wild salmon and steelhead runs in the Columbia-Snake watershed could recover, providing more salmon for other species, including the Southern Resident pods of orca living in the Salish Sea.

In the Southwest, water is scarce, and dam removal is being considered by some as a means of saving water. The Glen Canyon Dam impounds Colorado River water before it flows through the Grand Canyon. Less water travels down to refill it, and the water level in Lake Powell, the reservoir sitting behind the dam, continues to shrink. In 2023 this lake was at its lowest level since 1969. The hydroelectric capacity of the dam is severely reduced. More water is withdrawn than there is water flowing in. A white "bathtub" ring along the rock walls marks the prior water level. Less snow is falling in the headwaters of the river, and as the climate warms, the river loses more water from evaporation. Scientists estimate that for every 1.8°F rise in temperature, about 10 percent of the river's flow is lost.[13] Lake Powell has already dropped 170 feet. When it drops below a minimum power pool, the turbines will shut down. If the lake continues to drop below that, the dam will become a giant concrete plug in the river. Water consumption must be reduced. Farther downstream, beyond the Grand Canyon, is the massive Hoover Dam, with Lake Mead behind it. This reservoir currently sits at its lowest level in its history; in 2023 it was at 26 percent of its capacity.

The western United States has been in a drought since 2000, the worst mega-drought in twelve centuries. Scientists used tree-ring data to correlate the width of rings with soil moisture, a measure of drought. Human-caused climate warming has accelerated the drought. Colorado River water is severely over-allocated, diverted to water lawns in Arizona and Southern California and to irrigate millions of acres of crops in California's Imperial Valley and elsewhere in the arid West. About forty million people get their drinking water from this river.

Decommissioning the Glen Canyon Dam has become a goal for the Glen Canyon Institute. It asks, "Does a reservoir in the desert even make sense?" and maintains that by combining two reservoirs into one, millions of gallons of water would be saved.[14] By opening the floodgates of the Glen Canyon Dam, the water would flow down the Grand Canyon and be recaptured in Lake Mead. Less water would be lost to evaporation, and the hydroelectric power capacity of the Hoover Dam would be regained. This plan is not endorsed by the dam's managers, the Bureau of Reclamation. Still, if the dam were decommissioned, miles of river canyon would reopen for recreation and restoration. A free-flowing section of the Colorado River would benefit many threatened species, especially native salmon.

SALMON

To see for myself what good upstream salmon habitat looks like, I volunteer for a snorkel survey. I have been fitted out in a wetsuit, mask, and snorkel. As we stand in a group on the bank of a tributary of the Clackamas River, the fish biologist gives us some last-minute instructions. We are grouped into threes; I am to swim up the left bank, while Nancy takes the right bank and Brent takes the center of the water column. I am to count whatever passes through the water in the space from my left shoulder to the left stream bank. We are reminded to move slowly. We were trained how to identify coho and chinook salmon, steelhead, and trout while underwater, and how to estimate the size and number of schools of fish we encounter. We use our wrist-mounted pad to make tallies as we swim along, using special pens. The fish count will be used to compare the productivity in this stream with productivity in one that had been restored after logging.

Logging along salmon-bearing streams dealt another blow to salmon; poor practices led to erosion of forest soil and siltation of streams. In a healthy, intact forest ecosystem, trees and vegetation deflect the rainfall and water gets absorbed into layers of duff and organic soil layers. Where the forest is stripped away, rain falls on bare soil like tiny hammers, scattering soil with micro-ballistic force. As water flows overland and downslope on bare soil, it sweeps soil and rock along with it. As happened in Redwood Creek after its upper watershed was clear-cut, when upslope soil and rock wash down into the streambed, they fill the deep pools where salmon used to hide, and the spaces between the gravel where tiny eggs and developing salmon hide become clogged. The silted water absorbs more solar energy, warming, lowering its oxygen levels, and reducing survival rates of eggs and young fish. Logging along salmon-bearing streams now must leave buffers along the stream so that eroding soil becomes trapped there before washing into streams.

This stream where we are snorkeling has clear water flowing through rugged mountains covered with forest. The water feels cold on my face as I bend down to enter the stream. Beginning our slow crawl upriver, we use our hands to pull ourselves instead of swimming so as to be less likely to startle the fish. The light reflects onto the bottom substrate in a moving, reticulated pattern, like rippling giraffe skin. There are small pebbles, larger rocks, large boulders, and, farther upstream, some exposed bedrock. As we pull ourselves along, we notice caddis fly larvae in pebble-covered casings crawling over the rocks. Along the bottom of the stream are pockets of pea-sized gravel, preferred by adult salmon for laying their eggs.

We reach a logjam and pool; a group of small fish dart by. Each is 1–2 inches long, with a speckled back and a line of large vertical bars along its side, named parr marks. The parr marks are blobby, different from the narrow ones on a coho. I mark these as young chinook salmon and estimate forty in this school. During most of their first summer since hatching, chinook parr can be abundant here at the tails of a pool, where they eat the drifting zooplankton. As they grow, they will head to the mainstem of the Clackamas for deeper and faster water. They might stay there for a year, moving between rapids and pools while they turn a silver color. As smolt, they'll head downstream, swimming twenty-five miles of the polluted Willamette to the Columbia River, then one hundred more miles to the ocean. With so many predators and obstacles along the way, less than four out of this group of forty are likely make it to become adult salmon, I later learn.

The logjam consists of woody debris that fell into the stream. The wood becomes a substrate for insects, a shaded area where the water is cooler, and a safe refuge for fish to hide from predators or retreat out of swift channels. Young fish learn how to use the turbulence around the logjam to get a free ride. Deflecting the energy of the river, the logjam stabilizes the riverbank, and logs, branches, and sections of tree roots create a porous barrier that helps sieve material from the water.

While living in the ocean, adult salmon feed on marine plankton and small fish. Migrating adult salmon return upstream to spawn, dying afterward and then decomposing. Their bodies add nutrients back to the river and forest system. The marine-derived nitrogen from their decomposing carcasses provides thousands of pounds of nutrients; even juvenile salmon benefit from the nutrient enrichment.[15] More than forty species of animals—bears, birds, and arthropods—feed on the salmon carcasses and leave their scat and remaining bits of skeleton on the nearby forest floor, supplying it with nitrogen that increases plant growth. Tree-ring studies show that when salmon are abundant, trees grow up to three times as fast as when salmon are scarce.

Getting out of the stream, we group up to calculate our totals for the morning. Another team is counting fish in a stream that has been actively restored. Native trees were planted in the riparian zone, boulders placed, and more wood added to enhance the habitat for fish in the side channels. Later, the fisheries biologists will compare our data with that from the other teams to understand how restoration can improve stream conditions for salmon.

To mitigate salmon losses due to dams, fish hatcheries have been built and maintained along many dammed rivers. Instead of fish hatching out of the gravel in a clear headwater stream, they are hatched in a factory. Not subjected to the selective pressures of the stream environment where they might be eaten, hatchery fish are released directly into streams and can overwhelm wild fish populations through competition. Even though they have lower breeding success and lower survival than wild fish, they displace wild fish through their sheer numbers. They also interbreed with wild salmon, leading to a loss of reproductive fitness, loss of wild adaptive traits, and loss of diversity.[16] Advocates for wild salmon claim that the abundance of hatchery fish has led to greater declines in wild salmon. Although billions of dollars have been spent on hatcheries, the salmon population overall continues to decline. Despite the claims that we can have both dams and salmon, hatcheries have not contributed to an increase in salmon. Restoring the salmon's native habitat is a wiser solution.[17]

COLUMBIA RIVER

Roger and I travel to the lower Columbia River, where Chinook villages once lined the riverbanks all the way down to the Pacific Ocean. Anthropologist Jon Doenhke and Chinook tribal members have imagined a canoe voyage along the river by night five hundred years ago. From the canoe, hundreds of fires from these villages would have been visible. Thousands of Native people would have been living along the river. The Columbia River was used as a highway for thousands of years. The highway extended upriver to The Dalles, and downriver past the river mouth to travel up and down the coast. Canoes were important for Native communities throughout the Pacific Northwest, who used them to travel to visit each other for gatherings and for trade. Entire villages might travel together by canoe. The Chinook tribal population alone is estimated to have been fifteen thousand.[18] The Chinookan people of today say they have lived on the Columbia River for "time immemorial."

To make a canoe, a single western red cedar would be felled and carved out. The canoes had spiritual importance and were treated respectfully throughout the process of construction and use.

The Columbia River used to have many side channels and wetlands that acted as reservoirs during spring floods. These were productive areas for many foods the Native tribes depended on. A large and important Chinook village used to stand where the Ridgefield National Wildlife Refuge is now, downstream from Portland along the Columbia River. The refuge remains a productive area for wild food important to Pacific Northwest tribes because it still has acres of river floodplains and wetlands. We visit the refuge in spring to learn more about how Indigenous people tended and managed the native plants growing in the wetlands, sloughs, and meadows there.

JULIET McGRAW, an archaeologist, and her eight-year-old daughter lead the plant walk. Juliet is lively and bright, with dark hair hanging in braids, and wears a bright red jean jacket. Her daughter walks attentively by her side. Juliet explains how some of the knowledge the local tribes have is proprietary and sacred and kept only by the people. The floodplain was full of wapato, and the grasslands contained abundant elk and deer as well as plants for medicine and food, like camas. Western red cedar used for canoes and to build plank houses grew in the riparian forest. Different families and villages had rights to access these lands, which were controlled by complicated systems of kinship. The river and surrounding lands were sources of life for the people. Wapato, camas, and canoes were among the trade goods that Chinookan people had exchanged along the length of the Columbia River.

Like many other tribes, the Chinook were cultivators. It was women who managed and collected camas, wapato, and acorns. Beds of camas and wapato were maintained by processes of transplanting, weeding, burning, selective harvesting, altering the soil, and maintaining terraces. The knowledge about cultivation was passed on through generations of women. Juliet stops at a patch of camas, with its long fleshy leaves standing erect. Camas bulbs were collected in the spring and eaten after being steamed or roasted in underground pits for several days. Parched camas, ground into flour, was sometimes mixed with hazelnuts. Juliet explains how camas was also prepared for storage for winter food. The bulbs were slowly roasted over the fire and then compressed into cakes. After being hung to dry, the camas cakes were stored in boxes.

People would gather the camas in a way to ensure its future abundance. Camas fields were actively managed by periodic burning, releasing nutrients to the plants and discouraging competing plants. The bulbs managed in this way grew to the size of small potatoes. Other roots were also gathered—balsam root, edible thistle, fawn lily, wild onion, cattail, sego lily—and cooked by steaming in earth ovens before eating. Juliet shows us where tribal members still engage in wild tending; the leaves of plants competing with camas on a small rise have been braided together.

The wildlife refuge has areas of intensively managed seasonal and permanent wetlands. We follow our guides over one wetland where a swath of cattails grows along the bank of the large lake we pass. Their underwater rhizomes are thick and starchy. In spring, they would be collected, peeled, chopped, dried, and ground into flour. The rhizomes can be roasted and eaten whole. They have a mild buttery flavor like a potato. The stems could be pounded until soft and then used for diapers. The cattail leaves might be used in making baskets, rolled up to make a twine, or woven into dip nets for fishing. The long leaves might be sewn together and tied to wear around one's shoulders to serve as a raincoat.[19]

The wetland is where wapato—*Sagittaria latifolia*—grows. Hundreds of acres of wapato wetlands once existed along the lower Columbia valley. Explorers Lewis and Clark, who stopped here in 1805, named this area the "great water gardens of the Indians." Winter villages were located near these wetlands. Using special small canoes, women would paddle out into the shallow wetland lakes in spring to collect wapato with their paddles, or pluck them up using their toes while holding on to their small canoes. The tubers were cooked by steaming or baking in hot ashes. Settlers drained and plowed the beautiful camas fields and rich wapato marshes to make way for wheat, cattle, sheep, and potatoes. Here at the refuge, some areas of wetlands remain as wapato marshland.

Eating wapato is a healthy way to consume starch, because it releases its sugars in a sustained way. About a hundred years ago, diabetes was almost unknown to the tribes because they relied on native foods. But many had to switch from traditional cuisine to more accessible and cheaper fast foods. That, along with more sedentary lifestyles, has resulted in diabetes, obesity, and chronic conditions such as cardiovascular disease and hypertension among Native people.[20] Many other traditional Indigenous foods besides wapato are nutritionally superior, having lower levels of fat, sugar, and carbohydrates and higher levels of vitamins and minerals. Camas, for example, contains a prebiotic fiber called inulin, which helps control diabetes. Returning to native foods and ritual practices is now encouraged to bring healthy patterns back to tribal communities. The active lifestyle encouraged by traditional harvesting is also a positive health behavior, as well as being important to the people's cultural identity.[21]

In some of the upland areas of the refuge, low-intensity burning was used to manage oaks for acorn production, a process Juliet calls "tickling the roots." Burning the undergrowth using their ritual practices helped increase the acorn yield by reducing competition from other plants. It was women who tended the oak groves and grassy meadows, and the region's oak prairies existed mostly as a result of human modification. Hazelnuts were also managed using fire, which encouraged the shrub to grow abundant stems.

Cultural burning is one aspect of traditional ecological knowledge—TEK. Traditions like burning help conserve and enhance biodiversity, and using them would help us transition into a more sustainable way of living, says Linda Hogan.[22] A Chickasaw Indian, she writes about tribal thinkers who believe the natural world creates the human one: we are alive to processes within and without the self. She is one of many tribal people willing to demonstrate to non-Indians how to treat other living things as alive and how to respect the land so it can continue to bring forth sustenance for people. TEK knowledge was built over successive generations of experience, based on Native people's worldview and the strategies they developed for living sustainably.

Native people might explain that they first learned about the healing or nutritional properties of a plant through a dream. Vine Delo-ria Jr., a Sioux Indian author and activist, has explained how dreams help provide direct contact with the spirit world.[23] During dreams, people meet the spirits of birds and animals and receive special gifts of knowledge and power. Young people might need special training to become receptive to visions, but a vision could come to someone while swimming in a lake, a river, or ocean. Deloria described how a person talking with a plant was shown the plant's healing and nutri-tional properties. Instructions might be given by a butterfly perching on the shoulder of a young woman; the butterfly may instruct her in becoming a medicine woman.

OREGON TRIBES were forced off their homeland and moved onto res-ervations in the mid-1800s. Some were forced to travel far from where they had lived for many generations and move onto a reservation with people who had been their traditional enemies. In 1954, tribes were officially "terminated" and lost their reservation lands. In places where Native people had managed to keep hold of and own land, all services were cut off. Many people were forced to sell and move away. The fed-eral government helped private industries gain access to those Native lands to exploit their timber or minerals.[24] After protracted legal bat-tles, by 1980 five Oregon tribal groups were "recognized" again by the federal government, and some of their lands were returned. But the government still refuses to recognize the three-thousand-member Chinook nation as a sovereign entity.

Despite the centuries of cultural loss, traditional ecological knowl-edge has been maintained as families practice their traditions, use their recipes, and retell their stories. Applying this knowledge helps eco-systems recover and is important for Native peoples' cultural recov-ery.[25] Over the past thirty years, tribes like the Chinook have been making and using their canoes again. In an annual event, thousands of tribal members paddle their canoes along waterways to converge at a chosen Native center. The practices of canoe carving, relearning the dances and songs, how to navigate through the rivers and along the ocean shorelines, and rituals in how to come ashore are all being revitalized. The act of traveling together in the confined space of the canoe over long periods is part of working and respecting each other so as to rebuild a sense of community. Reviving cultural traditions like these are important to help tribes heal after more than 150 years of oppression and struggle.[26]

COLUMBIA RIVER GORGE

Driving back up the Columbia River, we pass the scenic area of the Columbia River Gorge, where the river cuts through the thickly forested Cascade Mountains. Moist air moves inland from the Pacific Ocean to here; it may rain as much as a hundred inches in a year. Dense forests of Douglas fir, hemlock, larch, and cedar cover the slopes. Dramatic waterfalls tumble over the lip of steep-walled basalt cliffs. The walls of the gorge were formed by a succession of twenty overland flood basalts of 10–15 million years ago, the same ones that created the high-columned cliff at Cape Lookout State Park.

A more recent cataclysm helped sculpt out the gorge. Glacial meltwaters from the last ice age used the river as a causeway to exit to the ocean. At several times, ice dams far to the east, temporarily holding back the meltwater, broke. The rushing waters flooded the Columbia basin with six hundred cubic miles of water traveling at about sixty miles per hour. Huge boulders were sprayed out where the dam broke; many more boulders along with tons of debris were carried as far as western Oregon over the two to three days it took for the water to flood to the ocean. These Missoula floods repeated over a hundred times between fifteen thousand and twelve thousand years ago.

We continue east and reach the dry side of the Oregon Cascades at The Dalles. We are now in the rain shadow of the Cascades, and brown grasses have replaced the green forests along the basalt cliffs. Powerline towers appear on the clifftops as we approach The Dalles Dam. Over The Dalles Bridge, we pass along the ugly utilitarian face of the dam itself, before reaching the opposite bank and turning northward to Horsethief Bluff.

We are here to visit the site where Five Mile Rapids used to exist. It was also called Coyote's Canyon or the Long Narrows by the Wishram Chinookan people. These rapids ended at Horsethief Bluff, river mile 194 near The Dalles. Although the word "Dalles" itself indicates there was once a gorge here, the city now sits alongside a flat reservoir lying behind the dam.

Before 1957, the river here had flowed in sets of rapids. One of these sets was called the Short Narrows, or Ten Mile Rapids; another was the Long Narrows, or Five Mile Rapids; and another was the famous Celilo Falls. The twisted textured basalt lying on the river bottom channeled the Columbia River into a narrow, swiftly flowing river. It was made by the same massive overland basalt flows of fifteen million years ago that formed the headland bluffs on the Oregon coast. People lived nearby and fished in these waters for over ten thousand years. The tribes left stone sculptures, mortars and pestles with carved ornamental designs, carved bone, nicely shaped charm stones, and ornamental beads of bone and stone that show how important this area was to them socially and for ceremonies. In a nearby canyon hundreds of petroglyphs and pictographs were created by Native people over thousands of years.[27] Some images are thought to represent the guardian spirits of shamans; some might have been created during vision quests.

Leaving our car at the parking area, we walk along a dirt path that circles Horsethief Bluff, a brown basalt pinnacle. The trail turns to go up the bluff. We slowly climb up to the crest of the butte. Below us stretches the flat, still lake of impounded water behind the dam.

This used to be the head of Five Mile Rapids, one of the best fisheries of the river. This was one of the first places the Chinookan peoples gathered to harvest the early spring Chinook salmon. Native people would await the salmon perched on scaffolds fastened to the basalt cliffs of the channel.

George Aguilar, a member of the Wasco tribe, remembered how the river looked back when he was a child. He described the churning, boiling river as it ran through the rapids.[28] Every bluff and hill had a legend associated with it. Rock formations, specific inlets, whirlpools in the river, and other geographic features were overlain with lessons and stories conveying to Native inhabitants information and expectations about behavior. These stories would be relayed to younger generations.[29]

The spring Chinook salmon would arrive by the thousands, heading up to their spawning ground. Aguilar described how people standing in their scaffolds along this stretch of the Columbia River skillfully caught large quantities of fish that sustained several thousand Native people during the winter. Thousands more from around the region came in the summer to fish and socialize and trade for dried salmon.

Five Mile Rapids, Ten Mile Rapids, and Celilo Falls were all drowned in the space of four hours when The Dalles River gates closed sixty-five years ago. As the water began to back up behind the gates, all the fishing stations were covered in the rising water. Then, "the river laid still and died," says Aguilar.[30]

One hundred fifty miles upstream from here, radioactive pollution from the decommissioned Hanford Nuclear Reservation still slowly leaks into the river. Pollution from factories and sewage treatment plants and polluted runoff from agricultural lands and city streets also enter the river. Columbia River tribal members who frequently catch and eat fish from the river are concerned about cancer risks. Fish like bass, sturgeon, carp, sucker, and perch that are resident and nonmigratory contain mercury and PCBs.[31]

Our modern systems have polluted and destroyed the life-giving capacities of this river. I recall how many people have worked to change salmon-killing dams. How do we keep from being complicit in the colonialism that helped destroy Native cultures? How do we make amends?

Today is Yom Kippur, the day when Jewish people fast, pray for forgiveness, and ask to be inscribed in the Book of Life. I feel grief over the fate of the river and of its people.

A small, dainty sparrow is perched in a shrub nearby. A Savannah sparrow. She opens her beige-colored bill and trills a few bars of song to me, startling me out of my reverie. I look over at her. She faces me. "Get on with it. Do something of value." Then she flies off.

"Inscribe all creatures in the Book of Life," I ask the sky.

CHAPTER 7

Wildlife

WATCHING KILLER WHALES swim by my kayak, I remember Paleolithic painted bison on the walls of Niaux Cave and sperm whale symbols carved onto Neolithic burial chamber capstones. We are bonded with animals by our experience and through culture. Even though our survival as a species depends on other living things, modern human activity now threatens one out of four species with extinction.[1] How can we repair our harmful ways? What if we paid better attention? More people would understand how animals link elements of their ecosystems; more would take responsible action on behalf of other animals.

WHALES

You can't help but be entranced by whales. Watching a whale breach brings a sense of wonder and motivates many to help protect whales and their ecosystem. Interacting with or observing whales in the wild, most people feel awe and a sense of kinship.

Orca, or killer whales, are top predators with incredible social capabilities. They have the largest global range of any marine mammal. Although we recognize their intelligence, their discernment is so different from ours it's hard to appreciate on its own terms.

Brad Hansen, a wildlife biologist, has been studying killer whales since 2005, when the Southern Resident group was listed as endangered. For Brad, whales' social affinity makes them seem more like us. Orcas live in groups—called pods—of up to forty to fifty animals. In the Northeast Pacific there are three distinct killer whale groups. Each prefers different prey, has different dialects, and a different social organization. Some say these whale groups each have a distinct culture. Each group is named for their movement patterns. There are resident whales, Bigg's or transient whales, and offshore orcas. Resident whales eat mostly salmon, while transient whales prefer marine mammals, especially seals and sea lions. Offshore pods appear to prefer sharks and other fish, following their prey twenty to thirty miles seaward.

Northeastern Pacific Resident orcas living around Vancouver Island and the Salish Sea are specialists, preferring salmon throughout the spring, summer, and fall. In one estimate, this whale population consumed as many as 820,000 salmon a year. Using echolocation, the whales can scan the air bladders of the salmon. Since each salmon species has a unique backscatter pattern, the whales can distinguish Chinook salmon from other salmon species.[2] This particular species of salmon is very fatty and nutritionally dense. Some scientists speculate that the killer whale's preference for them is cultural. Young whales are offered this fish by their mothers and can observe members of their pods actively hunting for these fish. The whales are selective; they will travel far from their "home" waters in search of chinook salmon instead of switching to a more locally plentiful type—pink, sockeye, chum, or coho.

Resident whales are easier to observe because they live and forage in inland waters. More is known about their life histories and behavior than about the other orca types. Their two distinct populations, Northern and Southern Resident pods, don't interact much and appear to use different vocal dialects. The Northern Residents spend their time around northern Vancouver Island and the central and northern coasts of British Columbia. The most studied resident killer whale population in the world is the Southern Resident, which spend their time around the inshore waters not too far from Seattle, Washington. They are a large, extended family that live in three matriarchal pods. They've been studied for the past forty years.

Brad explains how people recognize individual whales using photo ID. Each whale has distinct markings on their saddle patch below the dorsal fin. Using distinguishing markings, researchers recognized how the killer whale pods were organized by the matriarch. The three pods in the Southern Resident population are named J, K, and L. Male offspring stay with their mothers for their entire lives, seldom separating very far. As daughters reach sexual maturity at ten to fifteen years old, they might break off from the group.[3] Or they might stay with the group; there may be as many as four generations of whales living in a pod. During the summer months, the whales tend to concentrate in the Salish Sea, a region off the southern end of Vancouver Island near Haro Strait, the Strait of Juan de Fuca, and Georgia Strait. By looking at the genetic information collected from their feces, scientists learned only a few males fathered most of the calves. This means there is a high degree of inbreeding, which is not good for any population in the long run.

"The research is not like Jacques Cousteau," Brad remarks. "A lot of what we have learned is only indirectly observed." The matriarchs probably play a large role in guiding the pod; the matriarch of J2 pod was seen out in front, leading the group. Whale mothers share their prey with their own kin, but it's hard to know when this is happening. "We have to keep a distance from them while in a boat to avoid disturbing them," he says. Seen from the surface, their activity is subtle. The whales might be swimming a hundred or two hundred meters apart, and one will suddenly turn and swim to another whale and dive. Afterward, fish scales float nearby along with scraps of fish. Did they just share food? Other observers report seeing an adult whale release a salmon held in her mouth directly to a young one. In another example, a salmon captured by a whale was seen being moved to the surface, broken into pieces and pushed toward younger orcas. "Food sharing is a form of teaching, getting across to offspring this is what you do," Brad explains. Mothers begin to share less often with daughters when their daughters reach reproductive maturity.[4]

During their foraging/hunting dives, groups have been recorded underwater spreading out over a hundred or two hundred meters. One will dive and resurface, or small groups of whales will dive together and resurface with fish. They call to each other in a way that seems to coordinate their activities. The salmon have a low hearing sensitivity to the whale's vocalizations. Killer whales dive to the same depth as their prey and are known to swim along particular bottom contours.

Killer whales can recognize their own dialect; this enables them to reliably discriminate who is kin and who is not. For example, the stereotypical call of the K pod is a steady sound, while the J pod call makes a downward turn at the end, and the L pod call rises in tone at the end. Sometimes, a transient pod is recorded, and their calls sound more like a "woo" or can even sound like "hello," says Jeanne Hyde, a whale researcher. Repertoires of calls are like traditions; young whales learn the traditions of their pod.

Killer whale populations living in other parts of the ocean also hunt cooperatively and share their food. Off the coast of Norway, a cooperative feeding techniques called carousel hunting has been observed.[5] The whales herd herring into a tight ball, close to the surface. Schooling fish tend to form balls as a defense mechanism, and the whales make use of this, keeping the ball very dense and close to the surface by swimming around and under the fish while emitting large bubbles. One at a time, they stun their prey by slapping the edge of the school with the underside of their flukes, then eat the stunned fish. While some whales are herding and feeding, other whales swim around and under the ball of herring, lobtailing and porpoising. This behavior has been observed both from the surface and underwater. All the while, the whales were using echolocation. The sequence of call types seems to help coordinate group movements during carousel foraging.

BOB PITMAN, now working for the Marine Mammal Institute, has studied Antarctic killer whales for over twenty years. He identifies five different whale ecotypes or groups living there. Some avoid the ice altogether, foraging in the open ocean. Other groups forage among the pack ice for seals that haul out on the floes, using wave washing. Pitman has observed how they use cooperative behavior to wash seals from the ice floes. He explains how it works. The pack initially spreads out, spy hopping around the ice floe to observe if a seal is on top and which type of seal it is. The whole pod might spy hop around a floe to observe. They prefer Weddell seals to any other species. Once a decision is made, they bunch up, swim away from the floe, then charge at the ice together.

"They dive under the floe, creating a wave. The crest of the wave knocks the seal off the floe into the water. Then, a whale grabs the seal by its flippers and the food is shared among the pod. They almost always get their seal," Bob says admiringly.

Besides the wave washers and the open-water foragers, there are killer whales that forage along the leads that crack open in the pack ice and use those to hunt for fish. Pod elders appear to teach these methods to younger animals. These whales also learn which prey to catch through food sharing by the mother or other adults of their pod while they're young and by imitation. Killer whales have a remarkable ability to imitate; they can imitate on command the actions performed by another whale in an experimental setting.[6]

WE VISIT the Northern Resident pod of killer whales offshore of

Vancouver Island in Johnstone Strait, a channel between Vancouver Island and mainland British Columbia, Canada. This population numbers about two hundred and lives in large groups called clans. We board our kayaks and paddle along the shore, admiring the thickly treed hills on both sides of the channel. We scan the horizon for the whale's distinctive dorsal fin and whale blows.

It is a gorgeous summer day. The wind gently ruffles the water, and there is blue sky above. In the distance, a ripple on the water accompanied by an unexplainable vertical aberration of the air is heading toward us. Wait a minute, that's a whale's fin, I realize. The strange vertical aberration is the thin leading edge of the five-foot-high dorsal fin of a male killer whale. It slices through the water right toward us. He might be as long as twenty-eight feet and weigh as much as an elephant—five tons. Exhalations from the rest of the pod follow, sounding like *woosh*. As they dive, we can see the distinctive black-and-white markings on their saddles. This group of five is led by the matriarch, who might be as old as sixty.

Snapped into our little kayaks floating in the strait, we watch the pod pass us and continue south, heading toward the "rubbing" beaches where they will scrape themselves along pebbly rocks. When beach rubbing, whales skid their bodies over sloping beaches of smooth pebbles. Only one other pod of Alaskan resident killer whale is known to beach rub. This illustrates the differences in culture between populations of killer whales, says Jackie Hilderig.[7] The best-known rubbing beaches are on the northeast coast of Vancouver Island, along Johnstone Strait at Robson Bight.

We do not visit the rubbing beach out of respect for the whales' well-being; later we watch a video that was taken from the shore showing how the whales rub. So as to get down low enough to be in contact with the rocks, they super-deflate their lungs in a gush of bubbles to reduce their buoyancy. They rub all parts of their bodies along the pebbles. Sometimes they do this for a few minutes, sometimes for almost an hour. Beach-rubbing by the Northern Residents is assumed to be a social and recreational behavior, always done with members of their own pod. Vocalizations made by the Northern Residents while beach-rubbing are the same kind of vocalizations they make when their family reunites after a separation.

The Southern Residents, who live close to Puget Sound, are not doing as well as their northern relatives. Many whales appear under-nourished and thinner than normal. Many Southern Resident mothers have not been able to carry their calves to term, probably because of malnutrition. Observers used to see calves playing together, but in recent years very few calves have been seen. A killer whale mother was seen keeping her newborn calf at the surface even though it had been dead for hours. In 2020, the population was seventy-five, the lowest number in over thirty years. Fewer chinook salmon is only one stressor. Others are pollution from tire wear and persistent organic compounds like PCBs, which cause depression of their immune system when metabolized. Disturbance by vessel traffic and vessel noise interferes with whales' ability to communicate and hunt.[8] Collisions between whales and vessels are another problem. The agreement to clean urban stormwater from cities like Seattle and Portland might reduce pollution from the marine environment.

For people living near Seattle and Victoria, BC, killer whales are powerful symbols of wild nature because they survive despite their proximity to cities. Whales continue to symbolize the ecological preservation movement. Whales have a hold on us, some say, because as underwater dwellers they are symbols of the unconscious. Saving the whales is a way for us to "save ourselves from our own excesses, our own unrestrained efforts to tame nature," said psychologist Herbert Nieburg. When a whale is stranded, it's common for many people to show up to help: "it shows the human spirit as noble and heroic, like we would like it to be all of the time."[9]

Species can recover and rebound if serious conservation measures are taken. Because of protection afforded by the Endangered Species Act, humpbacked whales, gray whales, and some populations of fin and bowhead whales have increased in abundance over the past twenty years.[10] Vessels near whales are required to keep from striking whales, fishing gear has been modified to prevent many entanglements, and acoustical harm has been lessened by restricting military use of sonar and explosions near whales.

Killer whales are top predators that depend on many ecosystem factors, and saving them requires paying attention to these factors. Large mammal predators like killer whales are important to entire ecological communities. When their restoration is emphasized, ecological interactions are strengthened. Restoration of important river systems and

coastal and marine habitats throughout the Pacific Northwest is necessary to help endangered salmon, whales, and other animals. Restoring
rivers has a systemic effect on the watershed, extending to the coastline
and the ocean environment. Removing barriers to migrating salmon,
like dams that block their upstream migration to spawn, could help
these salmon-eating resident killer whale pods and other animals to
recover. Proponents of removal of the four lower Snake River dams
in the Columbia River watershed say this would give a big boost to
salmon populations.[11] To make up for the power that would be lost,
other dams can be modernized to generate more power.

JAMES NESTOR, who has taken up free diving, says humans possess,
but have forgotten how to use, a set of amphibious reflexes possessed
by whales, dolphins, and seals.[12] These reflexes—the mammalian dive
response—are learned and used by humans during free diving. During
the dive, heartbeat slows, blood moves to the more vital organs, the
lungs shrink in size, and divers can hold their breath for as long as
ten minutes. Free divers are silent and can go deep. One group of free
divers is learning about the language of sperm whales.

Working with acoustical physicists, free-diving photographers and
biologists in an organization called the Darewin Project swim with
sperm whales. They record the whale's vocalizations while observing and filming their underwater behavior.[13] The sperm whales swim
right up to them and began scanning them with sonar clicks. Diver
Fabrice Schnoller reports how the pulses of sperm whales' clicks are
so loud that you can feel them in your body. Looking for patterns in
sperm whale communication, researchers discovered discrete, repeat-
able, and complicated patterns of sonar clicks called codas. The unique
sequences of clicking sounds bond sperm whale clans. These are learned
and passed down through generations. Specific clans are named for
the special patterns of coda sounds they make. Sperm whales' complex
social behavior highlights their capacity for extreme intelligence. Other
researchers attempt to understand the patterns of sperm whale codas
by playing recordings of specific codas back to the whale, attempting
to communicate with them.[14]

The underwater photographer Fred Buyl said that when a whale
looks straight into your eyes, something changes in you. By inform-
ing others about whale intelligence, the team hopes to end commercial

whaling operations. Whales should have the same rights as humans, Schnoller maintains.

BUMBLEBEES

Bees play an outsized role in the ecosystem compared to their small size. By pollinating flowering plants, bees enable plants to develop fruit and seeds to propagate, and supply food and shelter for other creatures. Bees link many species.

In my own garden, many yellow-faced bumblebees (*Bombus vosnesenskii*) forage in the sunflowers and among the patch of lavender. Patterned with black and yellow hairs, they have yellow heads, a black thorax, and stripes of yellow toward the end of their body. This is a common summer bee in Oregon; most of the hive lives from April to September. Then, the newly mated queen abandons the nest and finds a place to hibernate over the winter beneath leaf litter or at the base of bunch grasses or in another sheltered location. She might nest in an abandoned rodent burrow, already insulated with fur.

We've placed small piles of brush and woody plant material around our garden, along with a compost heap, for nesting habitat for overwintering queen bumblebees. Other kinds of bumblebees who need a cavity in which to build their nest use hollow trees, abandoned bird nests, rock walls, or a hole under a tussock of grass. Queen bees are opportunists, looking for any suitably sized cavity.

Honeybees were imported from Europe in the 1600s. But they have been declining due to a host of factors, leading to colony collapse. Hundreds of other species of bees, like bumblebees, are native to Oregon and can help replace them as pollinators of crops. Studies show that pollination by bumblebees can produce bigger fruit and larger yields than from honeybee pollination.[15] Many native bees can forage even when it's cold and wet.

Bumblebees are cute; some call them the teddy bears of the bee world. They are easily told apart from other bees because they are large and fuzzy. Their fuzziness helps them carry pollen, which they can brush off using combs on their legs. They have phenomenal abilities to fly and sense the world. Their wings are flexible and interlocking. They can flap at more than two hundred strokes a minute, and can move up and down and forward and backward. Bumblebees often use buzz pollination, where the bee grabs a flower and vibrates her wing muscles to dislodge the pollen. But they are vulnerable to extinction. The Xerces Society, a nonprofit organization that helps protect invertebrates like bees, estimates over a quarter of all bumblebees are at risk of extinction because of loss of habitat, use of pesticides, and climate change.[16]

Bees can see into the ultraviolet end of the light spectrum. Many flowers have ultraviolet nectar guides, which, though invisible to our eyes, are to bees like runway lights at an airport, helping them find the center. Bees transfer pollen grains that became stuck to their "feet" and bodies from one flower to another. Thus cross-pollinated, those plants are more genetically diverse and potentially more adapted to a changing climate. Many types of plants will not bear fruit or develop seeds without pollination. Insect pollinators like bees also transfer microbes attached to the pollen grains that benefit plant growth and aid resistance to diseases.

"Bees are fascinating because they are all about interactions," says ecologist Andy Moldenke. While conducting a long-term study of native bees, he has come often to the wildflower meadows that dot the ridgetops in the Willamette National Forest. A group of us meet up in one wildflower meadow at Frissell Ridge, filled with a marvelous variety of wildflowers, with different shapes and colors: purple lupine, yellow monkeyflower, blue delphinium, and pink elephant's head. Andy shows us how to use the sweep net, deftly swooping the net over a bee on a wildflower, then twisting the net so the bee is caught. He carefully opens the net and fearlessly plucks out the tiny animal with his fingers. We gather round to look at *Bombus californicus*, a California bumblebee. We move on to another patch of flowers and group members try their skill at bee catching.

Over the morning, our group finds bumblebees with an orange band, sweat bees, long-horned bees, metallic green bees, and a fly that looks very much like a bee. All of the bee species found here do something slightly different, Andy tells us, yet their niches overlap considerably. Taken as a whole, bees are keystone species. They help connect the different wildflower meadows that exist along the different ridgetops along the spine of the mountains. Over time, Andy explains, we have seen different species decline and get replaced by other species. Bees need to move from one small meadow to another, and because the meadows are filled with diverse flowering plants, they can feed continually in the meadows. "Populations of each species are small, but every bee is moving all the time, and since there is so much diversity and redundancy, it creates greater stability overall for the ecosystem," says Andy.

Andy describes a study that revealed bumblebee intelligence. A group of scientists trained bees to use a tool, a small ball—training them to move it into a hole by rewarding them with sugar. Later, another group of bees watched these trained bees, and a third group watched a magnet drag the ball into a hole to get a sugar reward. All of the bees who had watched the first group of trained bees do their "demonstration" task caught on right away and could repeat the task. Bumblebees not only learned a new behavior, but also they learned this from watching other bees even though they themselves had no prior experience. Fewer of the bees in the group who watched the magnet move the ball could demonstrate the behavior. When the bees who had been trained were presented with a new challenge using three balls, and using different colored balls, most of them showed that they could solve the problem in a way different from how they had been taught. This demonstrated cognitive flexibility, say the researchers.[17]

A group of researchers in the United Kingdom devised a system of artificial flowers with wells of sugar water whose delivery was controlled by a string. The researchers used the same string-pulling experiment that had been used to study ape and bird intelligence. To get the sugar, the bumblebee had to tug the string. Some innovative bees solved the test right away. As in the ball-moving experiment, bees who had observed the trained bees at a distance also learned how to pull strings. When the colony could observe a live demonstration of a new behavior that a few members had quickly learned, the rest of the colony took much less time to learn the task and even improved on what they had observed. Researchers said this shows how bees have behavioral flexibility, beyond what had been inherited genetically. The researchers say these bees understand cause and effect and can use tools.[18]

Bees' antennae are extremely sensitive; they can sniff out good nectar sources, but they cannot detect pesticides. One class of insecticides known as neonicotinoids—neonics—are particularly harmful to bees.[19] The chemical is systemic, working its way into the pollen and nectar of the plants and persisting in the soil. These chemicals are highly toxic to bees because of their persistence, but appear to be less harmful than other pesticides to the humans who apply them. These chemicals are banned by the European Union; in the United States, the fight to restrict their use is ongoing.

As wild bumblebees decline, they are being reared, or encouraged to reproduce naturally, for agricultural purposes. Alfalfa farmers are learning to leave parts of their field untouched for alkali bees to nest. These desert-adapted bees of the western United States help pollinate alfalfa. So many millions of bees appear when alfalfa is in bloom that traffic through the area has to slow to ten miles an hour. Humans are now raising bumblebees to pollinate other crops, like tomatoes.

Because native bees are declining overall, people believed bees in the Willamette Valley and around Puget Sound were also in serious decline. After all, the Willamette Valley has changed considerably over the past two hundred years; prairies have been converted to farms and cities, and pesticide use is common. But because of a citizen science project, we now know many species of bees are still here. The Oregon Bee Project trained 250 volunteers to collect and help identify bees over two years. This data will serve as a baseline so we can compare bee populations in the future. Andy tells me a whopping five hundred species of bees were discovered living here, with a surprising number of bee species being found in more developed areas. It may be the result of all the restoration work being done by home gardeners and farmers. Andy reminds me that the best way to help native bees is habitat restoration to native plant conditions, and leaving space for them to nest.

HABITAT CONNECTIVITY

Connectivity between patches of habitat is important for bees. Bees can travel between gardens if they have opportunities to refuel often. Connecting patches of habitat is important for larger animals as well. Roads have fragmented what is left of wildlife habitat. By providing safe passage over or under roadways, animals can move around without the threat of being hit by a car or a truck.

One of the best-known sets of wildlife crossings is in Banff National Park, in Canada. Built when the Trans-Canada Highway was enlarged to four lanes, the overpass and underpass are used continually by wildlife. The best known of the passes is a grand-looking overpass entirely planted over with trees, native grasses, and wildflowers.

Grizzly bears, elk, moose, and deer are known to use this overpass. Smaller animals—wolverines, wolves, mountain sheep, red fox, hoary marmot, boreal toads, lynx, garter snakes, and beavers—use the underpasses. Before the wildlife crossings were built, the number of collisions between animals and vehicles was disturbingly high. Afterward, death of wildlife by collision dropped dramatically.[20]

In Wilsonville, fifteen miles south of Portland, a set of wildlife crossings was built under roads constructed over a wetland area. Beyond is a housing development with hundreds of closely placed single-family homes, but there are patches of open space and trails between groups of homes leading to a nearby regional park.

I meet up with ecologist Leslie Bliss-Ketchum, who has been monitoring how wildlife use these crossings for the past twelve years. Walking over to the first crossing structure, we reach the four-foot-high concrete culvert built into the side of the raised road and lined on the bottom with rock and soil. Stepping inside, I see tracks of coyotes, deer, raccoons, muskrats, and rabbits. This culvert connects the vast area of wetland with the remnant oak–Douglas fir patch on the other side of the road. Fences along the roads help direct the animals to the crossings. Leslie previously placed sand-track beds inside a larger culvert, and long-toed salamanders, frogs, and newts left tracks there. Animals you wouldn't expect so close to a housing development used another underpass for crossing. Prints of a short-tailed weasel were found inside another culvert. Preying on the abundant mice and voles, these weasels have appealing looks, with their long and lean bodies, perky faces, dark-brown fur, and white underparts.

Animals without safe passage over roads between their habitats are not as lucky. One recent winter night near Portland, people living near Forest Park found their road covered with crushed red-legged frogs. These amphibians were trying to get across two roads, a highway, and two sets of railroad tracks as they traveled from Forest Park to their breeding pond. These people joined together to manually help the frogs get across to their breeding ponds. Going out in shifts, all rainy night long, every night during the breeding season, the volunteers spread out, scanning the roadside for frogs. They captured the amphibians as they approached the first road, placed them in buckets, and transported them in cars over to the ponds. After they had successfully bred, the frogs attempted to cross the other way—and were transported back in a similar fashion to their habitat on the other side. The "frog taxi" service takes pride in having rescued 650 adult frogs last winter. They are working with different federal and state agencies to implement a future frog-crossing underpass.

HEADING HOME, I pass a dense housing development along Grahams Ferry Road. Abruptly, the houses end and farmland begins. Fields of blueberry bushes, row crops, and patches of forestland are interspersed with homes on large acreages. Continuing on, I pass the boundary into another city, and housing developments again appear. I continue on and return to the open lands of filbert orchards, marshlands, and forest. I have crossed the invisible line marking the urban growth boundary; we are back in rural farmland.

Preventing land from being developed in the first place is the best conservation strategy. Oregon has developed a set of unique land-use laws that give wildlife a better chance. Urban growth boundaries for every city were established, resulting in relatively compact cities and preserving open space. Both agricultural lands and forest lands were protected from development. By some estimates, Oregon's land-use planning has helped save 60 percent of farmland from development.[21]

Oregon's land-use laws were developed using the direct participation of Oregonians in the 1970s. Tens of thousands participated in drafting planning goals and guidelines during the hundred workshops held around the state. This built a wide constituency of voters, each with a personal stake in the program's success. At the time, a strong bipartisan coalition was able to help resist sprawl. Oregon has lost only a small percentage of open space to development when compared to states without these comprehensive land-use planning laws.[22]

BIRDS

Birds often use the wooded areas bordering streams—called riparian zones—to travel between forest patches. Tree swallows, willow flycatchers, yellow-breasted chats, and yellow warblers, who need to catch insects to feed their developing nestlings, are found in riparian zones in the spring and early summer in the Portland region. During early spring and fall migrations, these riparian areas are the stopover habitat for orange-crowned warblers and Wilson's warblers. Golden-crowned sparrows might overwinter there after migrating south from their northerly breeding grounds. Song sparrows live in riparian areas year-round. Predators, like Cooper's hawk, capture and eat songbirds there. Happily, many riparian areas are protected under Oregon's statewide land-use planning laws.

The larger the size of protected habitat patches, the better for animals and birds. Short-tailed weasels and the elusive northern flying squirrel both need large habitat patches, at least twenty-five acres. Many neotropical birds who migrate here for summer breeding, including colorful warblers, need a minimum of twenty-five acres of forest to feel comfortable nesting. Most Pacific Northwest warbler species benefit when large chunks of forest are preserved, restored, and interconnected via corridors.

Populations of birds can rebound when they and their habitat are protected. In a study comparing bird populations between 1970 and today, waterfowl populations showed a 56 percent increase and raptors a 200 percent increase because they have been protected using money set aside for conservation. But other bird populations in the United States and Canada have dropped by nearly 30 percent over the same period. Grassland bird species have declined the most; an estimated seven hundred million of these birds are now gone. The most likely reasons are habitat loss and pesticide use. "Every field that's plowed under, and every wetland area that's drained, you lose the birds living in that area," said one of the scientists in this study.[23]

PATCHES OF HABITAT can be re-created. In urban areas, even backyards can be restored to become tiny habitat patches for native birds. By maintaining their yard as a wild garden, planting thickets of native shrubs and small trees, and digging holes for water features, people re-create the native habitat on a small scale. Besides providing shelter, the trees and native shrubs provide food for insects like caterpillars, which in turn are fed on by birds in the spring. Other predatory insects and spiders help control the leaf-eating bugs. There are thousands of these yards in Portland, Oregon, and many more in places around the world.

A group of us studied the native habitat yards in Portland to understand how they benefited birds. More than twenty yards in three different neighborhoods were included in the study. Marilyn's house was a good example; it is located in a neighborhood with a good number of older street trees. She has a large Oregon white oak in her yard, with dozens of Oregon grape, snowberry, and red-flowering currant plants nearby. In general, the neighborhoods with the most tree cover had the largest number of birds and the greatest diversity of bird species. These yards also had the most insect prey. This was especially true where the canopy was made up of large leafy deciduous trees. This was where many insects and insect-eating bird species were counted. Out of all the species of birds, it was the black-capped chickadee that seemed to benefit the most from naturescaping.[24]

Michael Murphy—an ornithologist at Portland State University—and I stopped to watch a flock of black-capped chickadees that were hopping around and stopping to pick off bugs on the shrubs. With its distinctive black cap and bib and white cheeks and its perky nature, it is easily recognized. Michael explained how the "black-capped chickadee prefers smaller parks, edge areas, and broad-leafed-dominated forests. Habitat much like our backyards."

Chickadees eat animal food for most of the year: spiders, beetles, ants, caterpillars, and other insect larvae. Because of a specialization in their leg muscles, chickadees can eat while hanging upside down, so they can find food in places other birds cannot.[25] Chickadees nest in cavities, and might excavate an old cavity or start a new one in which to build their cup-shaped nest. Conserving dead trees—snags—is important to help chickadees. The female fills the cavity nest with feathers, animal hair, and down from plants. They mate for the season. In winter, chickadees might fly together in mixed-species flocks, with nuthatches, kinglets, and woodpeckers, collectively searching for insect prey or eating seeds. Chickadees can be surprisingly long lived; one black-capped chickadee that had been banded when young was at least twelve years old when it was recovered.[26]

Chickadees are pretty bold for their diminutive size. In his yard, Michael tells me, he can entice chickadees to eat sunflower seeds out of his hand. "Chickadees, more than any other species, can become comfortable with us." Michael remembers how he took his young granddaughter out for a walk when she was a little over a year old. Holding her in his arms, they spotted a chickadee in the bush. Michael proceeded to call it using a *pish* sound. The bird came closer, curious. Then his little granddaughter imitated the *pish* sound, and the bird moved to only a few feet away from them. Michael watched as the girl and chickadee stared intently at each other, mutually fascinated.

Chickadee calls are complex and language-like, communicating information on identity and recognition of other flocks as well as predator alarms and contact calls. The more "dee" notes in a "chickadee-dee-dee" call, the higher the threat level. Flocks have many calls with specific meanings. For researchers like University of Tennessee professor Todd Freeberg, the calls of the Carolina chickadee contain some of the characteristics of human language.[27]

Michael asks me what I think is the biggest threat to birds living in urban areas. "Flying into windows?" I guess. "No," Michael corrects, "it's outdoor domestic cats." Only about 50 percent of small resident songbirds survive long enough to make their own nests and raise a brood of young. "We need to ensure that populations of birds can successfully reproduce," Michael explains. "You might see a bird in your yard and think there are plenty of them, but it might be just passing through. By providing them with habitat acceptable to them to build a nest and reproduce, we can help wild birds."

Birds are experiencing a serious decline, and I ask Michael what we could do about that. "Millions of people love birds, spending more money than what is spent on spectator sport events to purchase bird-related travel, binoculars, and birdseed," Michael says. "If everyone who likes birds contributed only five dollars per year, we could help stem the decline of birds. We could generate a lot of resources to buy private land for bird habitat, restore corridors, and raise more awareness."

Given enough habitat, in parks and riparian areas, linked together by street trees and yards, and by keeping cats indoors, birds like chickadees can move and prosper even in an urban area. And if chickadees can prosper, so might other arboreal insect-eating birds like warblers, nuthatches, and woodpeckers.[28]

When people plant trees or shrubs and other plants in their gardens to encourage wildlife, they notice when birds and other animals show up. Paying attention like this generates a more emotional connection with the natural world. This seems to change how these people act to conserve broader facets of the natural world. Gardeners with an interest in caring for the wild birds also increase their appreciation for conservation of plants, insects, amphibians, and mammals, which are increasingly welcomed in gardens. And many of these wildlife gardeners become increasingly interested in other types of environmental issues.[29]

Hundreds of species of plants and animals are endangered and not protected under federal or state laws, according to scientists with the organization NatureServe.[30] Doing something to help protect biodiversity right where you live leads to protection for vital wildlife and plant species.

LEARNING THE LANGUAGE of the birds can reveal subtle signs of the outdoor life right outside our doors. Animal tracker Jon Young advises people who want to learn bird language to first allow nearby birds to become comfortable with their presence. Once you are less of a distraction for the birds, you will begin to notice what birds do and what sounds they make when they are going about their activities and when they become alarmed. Birds tailor their alarm calls to specific predators—like hawks or owls—and to the degree of threat posed. Every other bird in the area will respond to these calls. This is one of the ways a formerly hidden world can be slowly revealed to a prepared observer.

Naturalist David Abram reminds us that we don't discern the natural world with our intellect alone. He encourages us to "return to our senses" and let ourselves directly experience the living world. Intelligence is not just a human ability, he reminds us, but is shared by many species. When we pay attention to the lives of other animals, we become part of the more-than-human world. "We find ourselves living as members of a living, breathing, often suffering, body of relations."[31] Using senses beyond just the five common ones, our interactions with the natural world are enhanced.

Animals such as migrating birds and sea turtles use specialized senses to navigate. Scientists have not agreed on the exact mechanism by which they do this. According to John O'Keefe, May-Britt Moser, and Edvard Moser, we humans have specialized navigational cells in our brains that work like a GIS—geographic information system.[32] Polynesians might have used these when crossing the ocean without any modern navigation equipment. A recent sea voyage was made in a double-hulled canoe named *Hōkūle,* led by navigators Mau Piailug and Hawaiian Nainoa Thompson, sailing six thousand miles from Hawaii to Tahiti. The navigators say that besides their navigation skills, they used their instincts, and possibly magnetoreception. Their navigation skills included paying attention to the direction of the waves and patterns of clouds and the position of the sun and stars, and following cues from the flight pattern of fairy terns and noddy terns.[33]

CHAPTER 8

The Human Heart

"WE SEE THINGS, not as they are, but as we are ourselves," wrote Rabbi Shemuel ben Nachmani in the Talmud. If we see ourselves as separate, we isolate ourselves within our own skins. This perception of isolation is mirrored back onto nature. But when we see ourselves interconnected with other beings, we begin to understand ecology and global climate change and we care more for the living world. We begin to understand our place in everything.

ENVIRONMENTAL EDUCATION

Playing in natural areas, camping, walking, or collecting wild foods with their families, children develop an affinity for natural places and a feeling of kinship with other living things. Research has shown how these kinds of experiences during childhood can lead to a lifelong environmental consciousness.[1] Adults who are now environmental professionals often cite their childhood experiences in nature as pivotal to how they chose their life direction.[2]

I and a few friends organized a summer camp in the redwoods over five years. We joined with the children as they played among the old growth and searched for macroinvertebrates in Prairie Creek. We waded in the creek with dip nets and buckets and looked under rocks for stream macroinvertebrates: caddis fly, stone fly, and mayfly larvae. After placing their catch in buckets, we'd head over to the bank and examine these animals with hand lenses. The larvae are the immature forms of what will become the adult insects. The children learned to tell the flattened stone fly larvae, with tuft-like gills, from the mayfly nymphs, which scraped algae from underwater rocks. Crayfish were also captured. Tiny cutthroat trout were temporarily caught to examine too. The children returned the animals to the stream where they had found them.

Their favorite game was initiated at dusk. Donning dull-colored clothes, one group concealed themselves close to the trail and then the second group tried to find them. The "hiders" never entirely hid away behind anything. The seekers would slowly proceed down the same stretch of trail trying to find each hider. A seeker would identify who and where they spotted someone by calling out their name and location. Once spotted, the hider would stand up. After the seekers passed through, the remaining hiders would spring out, jubilantly yelling, "Ha! You passed right by me!" The game would be played until it was too dark to see anyone. We talked about how many animals and birds were cryptically colored and the behaviors they used to conceal themselves when necessary. The next night, many tried to behave in the ways of wild animals in hiding in plain sight. It was thrilling for them to crouch conspicuously yet remain hidden by the dusky light of twilight. Many felt as if they had become more like animals they admired.

We took longer hikes through the forest, played games centered around ecology. We told stories and sang together each night at our campfires. They'd crawl on hands and knees using a magnifying lens to examine the creatures living along a meter of ground on the forest floor. Many evenings, a herd of Roosevelt elk would troop right through our campsite, standing tall and majestically. We stood together and watched them with awe.

Over the years, the children were asked to keep a journal. Gabe wrote how he had watched the elk for over an hour without getting tired of it. Dominic liked the night hikes best, and was proud of how he could walk through the forest in the dark without falling. Heidi wrote that she could see all the wonderful things in nature that most people never noticed. Tanya wrote that she had experienced more at camp in a day than during an entire week at home. She felt good about how much she knew about stream life and enjoyed the attention she received because of that knowledge. Adria wrote about feeling like a different person while at camp, without the limitations she felt at home; she wrote that she liked the way she could express herself about nature.

These campers were fortunate to have families who valued spending time in the natural world and supported their child's nascent environmental awareness. The cumulative experiences they had in childhood supported their nascent environmental awareness. They grew to see themselves as people who knew about and cared for natural environments.[3] Four of the former campers, now adults, recently told me they still love spending time outdoors, now with their own families. One still remembers watching the baby frogs scatter as we hiked along the banks of the river, and scaring the bears who approached our camp one summer. Another remembers how she took a wrong turn on the trail and became lost while on a hike with the group. The memory of her success finding her own way back persisted. A third now works as a civil engineer for a city's water supply. The fourth is a musician whose inspiration comes from the natural world. He writes acoustical songs, including one titled "The Forest Through the Trees." They are all active advocates for environmental causes like mitigating climate change. I remember watching the wonder on their faces as we encountered the life of the forest, from the dragonflies to the old-growth trees.

OUTDOOR SCHOOL, where sixth graders spend a week at camp during the academic year, is another opportunity for transformation. "Intangible magic" is what happens for the children when they are here, said field instructor Vespa when I visited last spring. Far away from cell phones, media, family, and most social pressures, by the middle of the week, while the kids are busy doing outdoor science, a change can be seen.

"What did you notice?" the kids are constantly asked. The instructors listen to what they have to say. Often, the responses cascade over one another, one triggering others. This sharing of experiences with their camp counselors amplifies the impact of the child's stay in nature. It transfers to positive feelings for the natural world.

Walking back from their stream study, the students excitedly discussed ideas for a "time travel" skit for the evening's campfire. Later at the campfire, the students acted out the roles personifying animals and several nonliving parts of the stream—the tiny pebbles at the bottom, and the water itself—over different times over the year. Tiny salmon hatchlings pushed the pebbles aside as they emerged from the gravel into the water. Larger salmon smolts chased down aquatic insects to eat, and the water pushed the salmon downstream. The skit was accompanied by laughter from and recognition by their friends and teachers.

Spending time away from the constraints of parents or friends at home helps many children feel freer to express new aspects of themselves. For some, it's their first time away from home and a first time camping. It's the first time some of them have ever eaten fresh vegetables. Other kids come to Outdoor School already comfortable being outdoors. Learning how to get along with one another is a big priority. The staff intentionally mixes kids from different schools together to help build a sense of community. The instructors explain that kids who arrive at outdoor school surly and hard edged usually find themselves joining in.

After the morning's outdoor classes, I sit at lunch with Vespa, Captain Eco, Puddles, and Munk. The high school leaders and the adult field instructors all have assumed alias nature names. They explain how the kids, in the sixth grade, are the right age to learn about stewardship. Learning about forest animals is seen as important as the social and emotional learning about how to get along with one another. High school students volunteer as student leaders to work directly with younger students. Curiosity and interest about nature are communicated enthusiastically by the high school instructors.

Outdoor School helps increase a child's feelings of connection with nature. Prominent environmental leaders think so too; many of them remember their time in Outdoor School as revelatory. Staff members at Outdoor School choose to work here because of their own childhood experiences. "I remember how much Outdoor School inspired me when I was a kid," staff member Captain Eco reminisces. "When I was in high school, I came back seven times as a student leader." Sprout, another staff member, adds: "I liked it much better than school—in school, you read a book and write what you know. Here, it's all hands-on, every kid can shine."

Portland's Outdoor School program has taught over four hundred thousand Oregonians since it began fifty years ago. A legacy effect from three different generations of outdoor school attendees has created a groundswell of support for the program. There are outdoor school programs throughout Oregon. State lottery money is used so that every fifth or sixth grader in the entire state can attend Outdoor School.

One instructor described a fifth grader's pivotal encounter at Outdoor School. Sam had found a large, chunky brown "bug" clinging to a plant stalk next to the stream. Running down the animal's back was a slit containing a bright green bulge. The animal's shoulders—thorax—were poking out. Slowly, the head, legs, and the rest of the body emerged. An adult dragonfly. Sam was transfixed as the insect pumped fluids throughout its body and its wings expanded in size. The instructor explained that dragonflies were among the first insects to fly, 300 million years ago, and how some of them had enormous wingspans of nearly thirty inches. This dragonfly, a green darner, flew away, leaving behind an empty brown exoskeleton. Sam was wide-eyed from his contact with another living being, an encounter he would likely not forget.

Children may develop emotional affinities for natural areas where they spend time. And build self-confidence through outdoor challenge activities. Lessons that place a value on nature might lead to the child's future interest in learning about nature. These three things—emotional affinity, self-confidence, and nature knowledge—form a basis for future willingness to help protect the environment.[4]

CLIMATE CHANGE ACTIVISM

Jacob Lebel grew up spending time in forests and on farms. He has observed firsthand how the changing climate is altering conditions on his family's biodynamic Oregon farm. While trying to create a system of renewable energy and food sustainability, his practical research was challenged by long, hotter summers. His plans to help store carbon in the family's six hundred acres of Douglas fir were also being upset. They had thinned the forest to give space for other trees to grow larger. Jacob's dad used the harvested wood from the thinning operation to make furniture. They envisioned using a three-hundred-year rotation, eventually selectively harvesting some old growth. But the longer, warmer summers caused by climate change allowed beetle infestations in the forest. Trees were turning brown and dying. Widespread forest fires had begun earlier and earlier each season, with each successive fire season increasingly intense. Even though their forest had not burned, smoke filled the air and the sky turned orange. Everyone had to wear masks for weeks. Climate change is a lived experience for him and for other farmers.

Jacob became increasingly articulate through his time working in a coalition called Our Children's Trust. Working with climate scientists and lawyers, the group sued the federal government, demanding that it act now to protect our climate in trust for future generations. Their case—*Juliana v. United States*—was filed in Oregon's U.S. District

Court. Over the ensuing years, the group grew to include thousands of young people under twenty-five. Core members of the group lobby in Washington, DC, and hold press conferences to demand that those in positions of power act to help stabilize the climate. Their case against the federal government was dismissed in 2020, but they are seeking a rehearing. The group has legal action pending in seven states and proceedings in all fifty states to advance climate recovery. They participate in demonstrations, demanding action on behalf of younger generations, seeking a negotiated climate recovery plan.

Jacob tells me we have policies and plans already developed, but many of them are sitting on the shelf. "Put them into action!" insists Jacob.

COLLABORATION

Environmental writer Barry Lopez has written how many Americans are obsessed with personal gratification and operate with few social or environmental constraints. Lopez described how our dominant culture emphasizes a state of permanent adolescence. Many people in our culture demand immediate gratification, seek material acquisition and comforts, and glorify youthfulness. One anti-environmental group, Americans for Prosperity, fights regulations that would slow climate change, claiming to be protecting our freedom. Lopez thought we should grow up. He points to the restraint and respect for other beings evidenced by Indigenous people, many of whom befriended him during his travels.[5]

People working in forest collaboratives help resolve problems through consensus building and cooperative learning.[6] Many people in communities at the edge of national forest lands experienced years of adversarial relationships, particularly about their forest management preferences. Environmental groups had battled with logging companies, litigating over Forest Service management decisions they felt were harmful, while people who worked for logging companies protested about their need for a livelihood. While they were locked in standoffs, the forests grew more vulnerable to fire. Decades of fire suppression and increasing drought due to climate change were becoming serious threats to the forests.

Groups began working together—local landowners, members of environmental groups, the U.S. Forest Service, ecologists, small mill owners, and logging company representatives—to improve the health of the forests. Collaboration opened up the forest management process to more people. Collaborative members attend field trips and symposia held by forest scientists; they build bonds of trust over time. For people like Gail, an active member of the group, it was hard to accept forest thinning because it involves cutting trees. "It was very hard to say 'yes' to cutting down trees. I always loved trees," she told me. "But I observed how poor the trees looked, packed so closely together."

Controlled burns and manual thinning were used to remove many densely growing skinny trees and excess woody debris on the forest floor. This reduces the risk of a high-intensity fire by removing potential fuel. Both the forests and the communities have grown more resilient as a result.

An open, parklike look was already developing in a stand that had been thinned. Money earned selling the logs cut in the thinning paid for the forest restoration work. Gail described how, in one stand that had been thinned near Sisters, Oregon, a recent wildfire was much less intense and shorter in duration than it probably would have been if the forest had not been treated. Since there was little fuel left on the ground, the fire burned through quickly and caused less damage. The crowns of the forest were spared.

Gail explained: "As we met over time with people with different perspectives, we understood more about what the other side wanted. A collaborative mindset broadens your thinking."

Anybody can learn to cooperate. Cooperation occurs within groups of bacteria, colonies of bees, flocks of birds, pods of dolphins and whales, and among bands of humans. It was our penchant for social interactions that helped our large brains and intelligence to develop, says ecologist E. O. Wilson.[7] Bonds of mutual obligation is one explanation for humanity's success.

Over time, sharing decision-making authority and responsibilities among the stakeholders, private landowners, tribes, environmental activists, recreationists, local mill workers, timber companies, nonprofit organizations, and the Forest Service has helped build healthier forest conditions. But destructive wildfires that burned over five million acres in the Pacific Northwest in 2020–21 present challenges for forest collaboratives. So far, their sense of trust, network of social relationships, and mutual understanding of ecological problems continues.

Rehabilitating the burned-over lands will depend on future community engagement.

PAYING ATTENTION

We are conditioned to see ourselves as individuals, and as mere consumers. We might choose to buy ecofriendly products. We might feel so personally culpable for environmental problems that we carry the weight of the world on our shoulders. But we are not isolated individuals facing these issues alone. A recognition of our interconnectedness with other human beings, with other living beings, with the processes that sustain life on this planet, may start the kinds of changes needed in our Western culture.

I join the Global Climate Strike with ten thousand others in downtown Portland. The park and surrounding streets are packed with people. Most of the crowd is young-looking, college, high school, and younger students. A cheer spreads—we're jubilant over the size of the crowd, feeling positive about a potential shift in public opinion. Everywhere, people carry posters and signs. Some have a warning, some are funny, some are philosophical:

"You Will Die of Old Age, Your Child Will Die of Climate,"
"The Climate Is Changing. Why Aren't We?"
"Think or Swim."
"Make America Green Again!"
"If I Am Not for Others, Who Am I? If Not Now, When?"

Other protestors' signs remind us about solutions: "Renewable Energy," "Commute by Bike," "Build the Soil." Next to us, a young girl wears a sign saying "At Least You Could Try." Standing to one side, another group of silent people are wearing animal masks to represent wolves, bison, even domestic dogs. Calling themselves the Allied Megafauna, they all have a sign indicating they are "On Strike!"

We begin to march down the street. "Climate action! Now!" chants the crowd. As we march over the Hawthorne Bridge, below us a group of kayakers have rafted together and bear a sign aloft: "Climate Action!" In the water there is a group of twelve swimmers with small orange floats that bear the message "Swim with the Kids!" All the marching protestors create a line stretching a mile long, slowly crossing the bridge. The environmental movement of young people has been energized. These kids carry a moral weight—they have so much to lose if nothing much changes. Later we learn how four million people protested all around the world today. Will it make a difference?

Nonviolent strategies, like this march, can become powerful forces for change. Erica Chenoweth, a Harvard political scientist, found that civil disobedience, including nonviolent protests, was two times more successful in helping change politics than violent protests. She and her colleagues investigated 323 different campaigns and detailed the factors that cause some campaigns to succeed and others to fail.[8] Erica and coauthor Maria Stephan think the nonviolent strategies work better because they are supported by a wider swath of the population, and although they create a degree of civic disruption, these campaigns help form a more durable democracy. Eventually, they attract more people and erode the base of adversaries. Importantly, she says, once about 3.5 percent of the population becomes actively engaged in the movement, serious change begins. They cite the examples of the anti-apartheid movement in South Africa and the Arab Spring movement in Egypt against then-president Mubarak.

One group participating in the Climate Strike march were wearing scary Halloween-styled costumes with signs indicating they represented the CEOs of Chevron, Exxon, and Shell oil companies. Fossil fuel executives act as if we are unfettered by any need for restraint. The wealthy Koch family owns many companies including oil refineries and distribution systems. Besides allegedly spending millions to defeat mass transit projects, they spent billions to cast doubt about climate change by airing climate misinformation on public media, according to a Greenpeace investigative report.[9]

I MEET ACTIVIST Brian Ettling after the march. He describes how after reading Al Gore's book *Earth in the Balance* and seeing the documentary *An Inconvenient Truth*, he began to notice the impacts of climate change around him. Brian had been a park ranger in the Everglades, reveling in the abundant wildlife. Sea level was rapidly rising there, and he understood it could rise as much as three feet globally by the end of this century. Much of the Everglades is only three feet above sea level; he realized that many Everglades animals could lose their homes. It troubled him that some visitors would callously say to him, "Yeah, we wanted to see the Everglades before it's gone."

Brian explains how his activism grew. "If you love a place, and it is threatened, you want to help it." He had more to tell people, and felt confined to do so while wearing a Park Service uniform, so he became a private citizen. Brian signed up for The Climate Reality Project, for which Al Gore started using the proceeds from his Nobel Prize. Brian was trained to communicate more effectively about climate. He learned that lots of people are concerned about the situation, but were not sure about what to do next. He helps give them a voice. He uses humor to get his message across. "Humor helps people open up. It's hard to hate the message when you like the person."

Brian remembers the smoke that choked Portland and the Willamette Valley in 2022 from wildfires. Rising average temperatures are changing how and when snow and rain fall: we are having more rain and less snow in winter and less rainfall in the summer. Forests are drier in summer, a result of less summer rain and less mountain snowpack to melt. Wildfires in Oregon are also becoming more frequent and intense—and the costs to fight them are growing.

I ask Brian about the effectiveness of demonstrations and protests to get the government to act on climate change. "I have respect for people who do direct action; we have to stop the pipelines. But I consider myself an organizer. I'd rather work for change in the office of a congressperson, trying to find the sweet spot, where we have commonality." He is wary of cynicism. He tells me that activism is hard when compared with pessimism. Saying "Why bother?" is a way to rationalize your shortcomings. For Brian, activism is more rewarding. Protestors help put pressure on legislators, and so do personal letters and emails to them: Brian says we have to change the economic structure so that the business community gets behind renewable energy. But politicians need the political cover, they need to know that their constituents are behind something, because they are responding to so many issues at once. "When the public gets angry," he says, "that's when Congress makes deals."

CLIMATE CHANGE has accelerated the displacement of people vulnerable to flooding, prolonged droughts, desertification, environmental degradation, and sea-level rise. According to the Urban Institute, one million people in the United States have already left their homes for climate-related reasons: coastal flooding, drought, and wildfires. Worldwide, about twenty million flee their homes each year for climate-related reasons, according to the UN Refugee Agency. Slowing the progression of climate change becomes an imperative. But why has climate change continued unabated for so long?

Author Naomi Klein blames climate change on how we see the earth as a machine. This justifies the continued exploitation of the natural world for personal gain.[10] Carolyn Merchant says we began seeing nature without soul or purpose during the Industrial Revolution. Nature was just a clock, and animals were just automata that worked like clocks. Earlier biblical interpretations likened women to nature, ether virginal or disorderly and chaotic. Men were entitled patriarchs, heroes who brought order to the world, superior to all the rest. Men acted with impunity to take whatever they could.[11]

Modern society has based itself on a story of how perpetual material growth will bring abundance to all. But the story is hollow; natural resources are finite, the wealth is held by a few, and millions of people continue living in poverty. Economists like Tim Jackson suggest we should rethink prosperity—it shouldn't mean mere material success. If we live within ecological limits, people could flourish via social connections and well-being instead of material possessions.[12] The world's economy does not have to run on fossil fuel, because ample renewable or clean energy alternatives are available. Solar and wind power are becoming the largest types of new energy capacity being produced in Europe, and are almost 50 percent of the new energy in the United States. Electric vehicles are becoming more popular in a shift away from internal combustion engines. The transition could be financed by taxing CO_2 emissions, removing tax loopholes used by the rich, and phasing out subsidies paid to fossil fuel industries.

I WORRY if there is enough time for meaningful change to occur. But new ideas can spread rapidly. During the counterculture revolution of 1960–70s, ideas like equal rights for women spread rapidly and became widely accepted. Rachel Carson's book *Silent Spring*, about the dangers of pesticides, was published at the cusp of an awareness about the dangers of toxins such as those sprayed indiscriminately to kill pests. A new environmental movement quickly developed and spread. The rise in prominence of environmental justice began as part of the Civil Rights movement during the 1960s. Recently, awareness has grown about how disproportionally environmental degradation occurs where low-income communities of color live.

When residents in a low-income neighborhood in Portland, Oregon, learned how they had been breathing air contaminated with heavy metals like cadmium, arsenic, nickel, and lead, they joined with a group of lawyers, university professors, and political organizers to fight back. Forming the Neighbors for Clean Air, they lobbied state regulatory agencies for better air pollution rules. Due to their efforts and because of increased public support, homes in the area were fitted with air filtration systems, and stricter statewide air pollution rules were adopted.[13] Those heavy metals are dangerous to human health: cadmium damages the lungs, arsenic is carcinogenic, lead is a potent neurotoxin, and nickel can cause asthma and lung cancer.[14] This organization joined with twenty other organizations to clean up diesel pollution, which emits tiny lung-damaging particles, and create better statewide clean air standards.

EQUALITY

The United States has a very poor record of income inequality, ranking twenty-seventh among the richer countries of the world. Deprivations due to poverty lead to both health problems and intractable social problems. Researchers Kate Pickett and Richard Wilkinson compared rich countries in terms of particular health problems—life expectancy, infant mortality rates, mental illness, and obesity—and particular social problems—levels of homicides, imprisonment, and violence. They divided the countries between those with income equality and those with income inequality. In rich countries with more equal incomes, like Japan, Sweden, Norway, and Finland, levels of health and social problems are very low. However, the United States, with its high income inequality, has the highest levels of health and social problems among well-off countries. Economic growth does not make everyone better off unless there is more equality. Many modern societies, despite their overall affluence, are social failures, conclude the researchers.[15]

Societies where incomes are more equal tend to do a better job protecting the environment. According to another recent study, the countries with high scores for protecting biodiversity and habitat and environmental health also have high scores for income equality.[16] Meanwhile, income inequality is linked to biodiversity loss. In the United States, wealth has become more concentrated, while the number of threatened species has increased significantly and the number of individual animals and birds lost has increased.

Inequality is not a given—nor is equality a radical idea. We could rebuild our society to be more equal. Researcher Ingrid Robeyns proposes a new system she calls limitarianism, maintaining that no one should have over and above what one needs to lead a fully flourishing life.[17] The money saved could be used to help reduce our vulnerability to the effects of climate change. We could invest in drawing down and storing greenhouse gases by protecting existing forests, growing new forests, and increasing reliance on renewable energy and energy efficiency. If protected, existing forests could store between 5.5 and 8.75 gigatons of CO_2 over the next thirty years—one gigaton is one billion tons. Growing new forests can draw down as much as 3 gigatons of CO_2 each year. We could avoid adding more carbon to the atmosphere by increasing reliance on renewable energy and energy-efficient technologies. The wind energy potential of Kansas, Texas, and North Dakota alone could supply electric energy for all of the United States.[18]

Portland activist Betty Shelly believes American-style overconsumption makes all our environmental problems worse and increases inequality. When she learned more about the environmental costs of making the products she was buying, she resolved to live more simply. "Everything we buy has an impact," she told me. It takes fossil fuel to make a plastic water bottle, so she felt it was immoral to even buy a throwaway product like that. She began to collect waste water, made a compost pile, recycled everything possible, and bought less. She and her family became so frugal that they produced only one can of garbage a year. Working at a clothing store, she observed how the certain women came to the store frequently to buy an item. The item they bought last week apparently no longer gave them pleasure. Later, she noticed many of those women were actually in the mall every day, always shopping for something new. Betty understood that shopping was not a means to leading a satisfying life. What satisfies us is the quality of family life, meaningful friendships, meaningful work, leisure time to develop one's own talents, says Michael Argyle in *The Psychology of Happiness*.[19]

Betty began to teach a class to anyone interested in how to live more sustainably and satisfyingly. "It felt really good to try to help things change," she reminisced. Taking her ideas from Jim Merkel's *Radical Simplicity*, she asks the students to imagine all of the world's natural resources displayed before them like at a potluck, with all the world's necessities for life, all of the clean air, water, forests, food, and soil there.[20] "Imagine Americans as the first in line at the potluck. What will you heap on your plate? How much is enough to leave for your neighbors behind you in the line?" She asks the students to picture all the rest of the world's population standing in line, standing shoulder to shoulder, waiting behind Americans at that buffet. And along with them are all of the "millions of other animal and plant species, millions of billions of unique beings, all with the same lusty appetites. And behind them, future children, lion cubs, whale calves, and even insect larvae." Looking at the world's resources this way requires us to use restraint. What is the level of equity that we would feel good about? she asks her students. At what level of inequity would we say, "Wait a minute, that's not fair."

ACTIVISM

Some of us become active environmentalists. Mike Houck launched his forty years of activism by writing letters to urge the Portland city government to protect wildlife and later by conducting fish and wildlife inventories. "At the time, there were only maybe six of us who believed nature needed protection in cities," he reflected. "I was told by then local planners that there was no place for nature in the cities, that cities were dead zones." Mike was designated the first urban naturalist for Portland Audubon. Since then, attitudes toward nature in the city have changed, and hundreds of thousands of people all around the world are engaged in restoration of urban ecology.

Early on, Mike studied Portland's regional wildlife system. How much land was already protected, and where was it located? He recognized, along with others, that some of the habitat they thought was protected was actually not, and so was being lost through development. Mike worked with Metro, the regional government for the Oregon portion of the Portland metropolitan area, to take aerial photos of the three-thousand-square-mile Portland-Vancouver region. Together with a team from Portland State University, he assembled the photos into a large map showing what was "green" and what was not. Using the maps as evidence, along with his formidable interpersonal skills, he helped build a coalition of organizations and advocates that convinced Metro that greenspace was important to protect. Metro began passing bond measures to enable the purchase of more land for preservation and recreation. Now, over seventeen thousand acres of open space are protected. He worked to elevate the issue of habitat connectivity via urban streams and to promote affordable housing through his work with the Coalition for a Livable Future.

Protecting natural habitats and biodiversity is even more significant because it can protect human health. Deforestation and capturing and selling live wild animals expose people to new infectious diseases, especially coronaviruses, Mike reminded me recently.

We sat together in his backyard; a chickadee called *dee dee dee*. "Aw," he sighed, "I love that bird." Mike recalled how, when prime bird habitat near downtown Portland was proposed for development, he and a friend created forty signs reading, "Wildlife Refuge/City Park." One night, they went to the bottomlands and posted all of the signs, high enough so that their removal would be difficult. Within two weeks, the media was talking about the new city wildlife refuge. Within a few years, Portland adopted its first urban wildlife refuge. Since then, birders have seen over 180 bird species at the refuge, and many hikers and joggers visit and kayakers paddle by. "Sometimes, you have to create your own reality," he grinned.

IF ACTIVISM is limited to confrontation and demonstrations, even if well intended, it results in reactivity and even suffering. Activism needs to be more systematic to be effective. David Peat proposes we engage in "gentle action," acts of "creative suspension" or "alert watchfulness," before engaging in more active or forceful interventions.[21] Although we usually react without thinking, if we reflected first, we might respond more authentically. Reflection can give us greater freedom over external forces and over habitual prejudices.

David Nicol thinks we might help heal some of the "planetary shadow" from the legacy of colonialism and industrial capitalism by making ourselves aware of the unstated assumptions, wounds, and limiting thought patterns we inherited. We might keep from reproducing our problems all over the planet.[22] Nicol advocates for more subtle forms of activism. Subtle activism involves individual inner work through meditation, prayer, or other spiritual practice, and a transformation of our limiting beliefs and behavior patterns. We can identify a vision of what we hope can happen, working with groups to build a sense of connectivity.

Appreciating our connectivity is important. In one study, people with more collective orientations tended to feel that their smaller individual actions would, when taken together, have an impact. They were more likely to do something about climate change. Individualist thinking causes us to become more alienated and intractable. Participants in the study with individualist orientations saw climate change as intractable; they were not motivated to do anything about it.[23]

Nearly fifty years ago, professor Christopher Stone proposed that trees, animals, and natural entities like rivers, bogs, and deserts should have a legal standing in the courts.[24] By changing how legal rights are ascribed, existing environmental laws can be reformed, and laws that legalize the extractive economy can be challenged. Now, laws protecting nature as a legal entity have been established in Ecuador, Bolivia, Uganda, Australia, and New Zealand. In the United States, organizations in Orange County, Florida, legalized the protection of two rivers, and recognized their right to have healthy ecosystems.[25] Groups in Rhode Island spoke up on behalf of their four hundred miles of coastline in a court case about climate change. Their case accused fourteen fossil-fuel industries of covering up the impacts of fossil-fuel use on climate change. Commercial fishermen in Oregon and California sued thirty fossil-fuel industries for causing the rise in greenhouse gases, leading to ocean acidification, rises in algae blooms, and increases in neurotoxins. As a result, crab, shellfish, and fish populations are declining. The general public should not be on the hook; the harm should be covered financially by those who perpetuated climate change, maintains Noah Oppenheim, of the Pacific Coast Federation of Fishermen's Associations.[26]

This legal movement helps popularize a worldview already widespread among Indigenous peoples, that nature is a living being with rights to exist as it is. The Ojibwe tribe in Minnesota protected the rights of a staple food in their diet, wild rice, or manomin, to clean freshwater resources. The Whanganui Iwi group of tribes in New Zealand worked for years to protect the rights of one particular river. It finally was granted the legal status of a person under the name Te Awa Tupua. The river has the same legal rights and responsibilities as people and corporations. "We consider the river an ancestor and always have," said Gerrard Albert, a tribal member. "Rather than us being masters of the natural world, we are part of it."[27]

THE TREE OF LIFE

We also have inheritance from common ancestors of whales and otters, ravens and marbled murrelets, redwood trees, and single-celled archaebacteria. Charles Darwin once depicted evolution as a tree of life that connects all species, extinct and living. The Tree of Life that grew in the Garden of Eden is part of the Judeo-Christian origin story. Many cultures highlight trees as symbols in their own creation stories. Ancient Babylonians believed a cedar tree grew where humans were first made out of clay by the Sumerian god Enki. The Sumerian creation story describes a tree growing in the navel or center of the world. Eating of it would make one immortal. The sacred tree evolved in Western culture as a symbol of self-renewing life. In Europe, folk cultures honored the May Queen through dancing around maypoles decorated to celebrate the sacred tree and help encourage the spring renewal. Non-Western cultures, such as Indigenous Potawatomi people, have a creation story involving the great Tree of Peace. Hopi legends include a great tree living underground in the place of humanity's emergence.

Many cultures used to believe the earth itself was a nurturing, living organism. For peasants living in Europe during the Middle Ages, every tree, every spring, stream, and hill had its own guardian spirit. Before one cut a tree, mined a mountain, or dammed a brook, it was important to placate the spirit in charge, and to keep it placated. The belief in spirit-guardians constrained deforestation, pollution, mining, and overall destruction for long periods. By the seventeenth century, those beliefs were replaced in Western culture by more mechanistic beliefs about the earth, leading to greater indifference to the feelings of natural objects. According to Lynn White, professor of medieval history, as older cultural inhibitions on exploitation were destroyed, religion became a means for humans to feel superior to and even contemptuous of nature.[28] Modern Western attitudes of entitlement and individualism are rooted in the Renaissance, the Enlightenment, the Romantic movement, colonialism, and a global market economy.

There are still places where environmental destruction is prohibited and traditional customs hold sway. Older customs restrict and prohibit destructive activities like hunting, grazing, and logging in these places. Around the world, sacred sites are protected by laws and taboos. Sacred groves are preserved in India, in mountain regions in Mongolia, Japanese temples, and in forests and temples of Cambodia. Sacred forests in Lebanon, landscapes with groves of trees in China, and sacred gardens in Borneo continue to be preserved. Traditional stewardship practices have helped manage these sacred places. Many of these sites protect wildlife living there; wildlife like the mountain lion are also important to maintain biodiversity.

In one of Hayao Miyazaki's animated films, *My Neighbor Totoro,* two young girls have moments of communion with nature that echo Japanese Shinto spiritualism. At the beginning of the movie, their father pays respects to the spirit of the giant tree. He describes how the ancient tree existed there from a time when trees and people were friends.[29] Shinto, an ancient nature-centered spirituality emphasizing an intuitive continuity with the natural world, is experiencing a revitalization in Japan. The resurgence of Shinto is linked to forest conservation; and sacred forest sites are protected, ranging from small groves of trees to large areas in Japan. Once protected from exploitation with taboos, now many are protected by the vigilance of activists and through conservation initiatives.[30]

Traditional societies that are successful in maintaining their forests, wildlife, and wetlands have embedded sustainable practices into religion and rituals. Tribes living along the Columbia River constrained their behavior toward when and how much salmon was harvested. As salmon began to run upstream, no fishing was allowed until after weeks of ceremonies were completed. Practices might regulate more egoistic and wasteful behavior and embed the message of responsibility.[31]

Particular rituals continue to knit people with nature. Through rituals, people acknowledge their link with ancestors, with their children's future children, and with other living things. Every year during changes of the seasons, people in some rural communities in Europe participate in ancient festivals, such as the Bulgarian Kukeri ritual. In ritual masks and costumes, people parade through the crowd wearing clanging bells to help restore the fertility of people, animals, and agriculture. Rebirth ceremonies are held from December through spring in which men and women dress up as animals, stags, "beasts," or as wild men. These rituals fall during a time of great cold, ice, and snow, but life begins as the hours of daylight begin to increase. They may stir a sense of the sacred, elevate living things as rare and precious, inspire feelings of kinship and love, and demand devotion in the form of actions. During rituals, people feel a part of something larger than themselves.

In Romania, men dressed as stags portray a hunt. L'homme sauvage dressed as a bear waking from hibernation in the Pyrenees helps banish winter; a wild man dressed as a mythical beast with horns and long teeth may create mock terror in the crowd. A wild man can be dressed in lichen, leaves, tree branches, or straw, completely covered from head to toe. Whether dressed as a stag with a bright red cloth portraying resurrection after being slain, or wearing a costume covered in thousands of burrs known as "Burryman" in England, modern Europeans reenact what was once a sorcerer's role in a human-animal transmogrification.[32]

A costume once worn during a ritual dance in the Kukeri festival in Bulgaria is displayed at the Musée de l'Homme, the French anthropology museum. I examined these costumes up close during a visit to Paris. This costume consists of a grasslike green cape, a belt of huge cowbells, and a mask covered with mosaic tiles. Red beads surround the eye holes, and the long nose is birdlike. Nine fantastical man-animal masks are also displayed. One mask is a horned, fur-lined snarling animal face; there is also a goat-man mask complete with curving horns and a strong wooden-carved face, and there are comical clownlike faces and scary witchlike masks. They are sometimes sewn with sequins, beads, and shards of mirror, the colors themselves evoking some ancient symbols of the forces of nature: red for sun-fire and fertility; black as Mother Earth; and white for the pure, life-giving energies of light and water.

The Green Man is a sculptured face carved into the stone lintels of churches and other buildings. He has leaves erupting out of his mouth, nostrils, eyes, and surrounding his face. This image was common in buildings built throughout Europe, Lebanon, and Iraq

during a medieval period of resurgence of paganism. The Green Man
still retains some life-renewal symbolism, indicating resilience.

IN MID-JULY, Roger and I attend an annual ritual celebration of paganism in Oregon. We drive south of Portland, park near Eugene, and hop
on a bus that takes us to the Oregon Country Fair. Started in 1969 as
a fundraiser for an alternative school, it recently celebrated fifty years
of continued countercultural immersion and celebration. We find meditation, relaxation, and reconciliation tents set up under structures
made of branches, draped with colorful cloths, and with rugs and cush-
ions spread around on the earthen floor for comfort. There's a parade
down the path, made of a huddle of flower-wreathed costumed flower
people: men covered from head to foot in green leaves or colorful flowers, women in floral headdresses and colored skirts sewn with flowers,
and a man with his entire chest and head covered in leaves and flowers, with the botanical detail of one of Giuseppe Arcimboldo's fruit-
man portraits. Diversity is celebrated. A merry man clad in pink and
red flowers grins impishly as he prances by. Most of our own group,
by contrast, are comparatively drab and bland-looking in our con-
formist T-shirts and shorts. Everywhere, musicians play bluegrass, jazz,
folk, mandolin, guitar, flute. Paleolithic-styled men, wearing animal
skins, demonstrate the miracle of making fire using the ancient drill
method, and flint knappers show us how to create stone tools. Earth
mothers, draped in green veils with faces painted silver and sequined
with zigzags, glide past. Men in tutus, stilt walkers dressed as fan-
ciful creatures, gymnasts with feather and flower headdresses twisting and posing from circus ropes also pass. Another parade passes:
there's a massive pterodactyl-headed puppet person draped in shim-
mery scarves, a giant papier-mâché queen of the salmon spirit draped
in blues and greens, powered by a puppeteer, followed by a school of
papier-mâché salmon held aloft by participants trailing behind, followed by a dancing crowd of festival volunteers in colorful skirts. A
mock village seemingly from Renaissance times has been constructed
of cedar shakes, with stores selling crafts and food on the ground level,
while from the second floor the families of temporary shopkeepers peer
down and watch the parade of people. Some storefronts are fancifully
built; one resembles a huge raven's head. Music pervades the air. The
smells of cooking food and the sounds of laughing and dancing people

are everywhere. Paganism is celebrated here as modern ritual tradition. People tell us they are celebrating and expressing reverence for the earth through dancing, drumming, singing, and chanting.

LIVING WITH THE WILD

Our current economic and social systems are unraveling as the earth steadily heats up. Species continue to disappear, and life-providing ecosystems deteriorate. We need to help each other out and work with our local community to support local food production, emergency preparedness, and local governance to build resilience.

Cleaning up one's own place is the least I can do. Roger and I held a world renewal ceremony on our land. Roger dipped the roots of the trees we were about to plant into a solution of mycorrhizal fungi, and together we planted the first tree with ceremony. The previous landowner had cleared the forest to plant an orchard; we replanted the forest, setting out 120 small Douglas fir and ponderosa pine trees, dozens of young snowberry, ocean spray, and flowering red currant shrubs. As they grow, they will take up increasing amounts of atmospheric CO_2 and build more habitat for invertebrates, small animals, and birds. Standing together, we acknowledged that we face the possible extinction of our species, along with so many others. As we planted, we felt a deeper connection to this land. We grieved over what is being lost, and vowed to work together to protect what we can. Today, some of those trees stand over six feet tall.

Why do most of us feel powerless to change things? It's because we allow ourselves to feel that way, says Frances Moore Lappé.[33] Individualistic thinking contributes to the growing dread many people feel. By actively working together, our actions will feel more meaningful.

How might we learn to live within natural constraints? Although we recognize our potential for acting selfishly, we try to not give into it. By remaining in connection with what is being lost, we behave as if everything we do matters. Acknowledging that we share that world with other beings, we open our eyes to the beauty and the precariousness of our situation. By doing something of value, we can help enrich the world.[34]

Notes

CHAPTER 1. THE REDWOODS

1. Flannery, *Future Eaters.*
2. See https://www.savetheredwoods.org/about-us/mission-history/redwoods-timeline/.
3. Speece, *Defending Giants.*
4. Grosvenor, "World's Tallest Tree."
5. Agee, "Issues and Impacts."
6. Save the Redwoods League, https://www.savetheredwoods.org/about-us/mission-history/redwoods-timeline.
7. Sealy, Guerra, and Amodio, "Redwood National Park Expansion."
8. Hudson, "Sierra Club v. Department of Interior."
9. Merhaut, "How Do Large Trees."
10. Ewing, "Fog Water."
11. Bakeless, *America as Seen,* 311.
12. Meany, "Diary of Wilkes in the Northwest."
13. Binkley et al., "Role of Old-Growth Forests."
14. Morrison and Swanson, *Fire History and Pattern.*
15. Egan, *Big Burn.*
16. "Giant Logged Long Ago."
17. Flavelle and Fountain, "Oregon Burns."
18. Crotteau and Keyes, "Restoration Treatments."
19. Bürgi, Östlund, and Mladenoff, "Legacy Effects."
20. Friedman and Reich, "Regional Legacies of Logging."
21. Dupouey et al., "Irreversible Impact."
22. Isbell et al., "Deficits of Biodiversity."
23. Wohlleben, *Hidden Life of Trees,* 57.
24. Prinsloo, "Mycorrhizal Fungi."
25. Simard, "How Trees Talk."
26. Simard, *Finding the Mother Tree.*
27. Meigs et al., "Influence of Topography."
28. Lutz et al., "Global Importance."
29. Gray and Whittier, "Carbon Stocks and Changes."
30. Law et al., "Land Use Strategies."
31. Hart and Sailor, "Quantifying the Influence."
32. Tushingham, *Archaeology, Ethnography.*

CHAPTER 2. THE PAINTED CAVES

1. Lewis, "Cascadia Cave."
2. Minc and Smith, "Spirit of Survival."
3. Hassett, *Built on Bones.*
4. Tattersall, *Masters of the Planet.*
5. Sorenson, "Pre-Conquest Consciousness."
6. Sauter et al., "Cross-Cultural Recognition."
7. Lewis-Williams, *Conceiving God.*
8. Chauvet, Brunel Deschamps, and Hillaire, *La Grotte Chauveta,* 42.
9. Von Petzinger, *First Signs.*
10. Utrilla et al., "Paleolithic Map."
11. D'Huy and Le Quellec, "Les animaux 'fléchés.'"
12. Marshack, "Taï Plaque."
13. Groenen, "Images de mains."
14. Broome, *Aboriginal Australians.*
15. Lewis-Williams, *Mind in the Cave.*
16. Clottes and Lewis-Williams, *Shamans of Prehistory.*
17. Raux, "De la grotte ornée."
18. Delage, "Comments on a Recent Challenge."
19. Fuentas, "Depiction of the Individual."
20. Sahlins, *Stone Age Economics.*
21. Kahn, "Foragers to Farmers."
22. Flannery and Marcus, *Creation of Inequality.*
23. Otterbein, "Archaeology of War."
24. Lipson, Szecsenyi-Nagy, and Reich, "Parallel Paleogenomic Transects."
25. George, "Code Hidden in Stone Age Art."
26. Clottes, *What Is Paleolithic Art?*
27. D'Errico and Henshilwood, "Origin of Symbolically Mediated Behavior."
28. Goebel, Waters, and O'Rourke, "Late Pleistocene Dispersal."
29. Rutherford, *Brief History of Everyone Who Ever Lived.*
30. Shillito et al., "Pre-Clovis Occupation."
31. Fleur, "Earliest Known Human Footprints."
32. Cosmides and Tooby, "Evolutionary Psychology."
33. Barnes, *Rewild Yourself.*

CHAPTER 3. STANDING STONES

1. Zeder, "Origins of Agriculture."
2. Harari, *Sapiens.*
3. Hassett, *Built on Bones.*
4. Manning, *Against the Grain.*
5. Cortazar, "Scholar: Cave Paintings."
6. Curry, "First Europeans."
7. Schulz Paulsson, "Radiocarbon Dates."
8. Lewis-Williams and Pearce, *Inside the Neolithic Mind.*

9. Cummings, Midgley, and Scarre, "Chambered Tombs."
10. Scarre, *Monuments and Landscapes.*
11. Lewis-Williams and Pearce, *Inside the Neolithic Mind.*
12. Whittle, "Very Like a Whale."
13. Scarre, *Monuments and Landscapes.*
14. Richards et al., "Settlement Duration."
15. "Orkney Venus."
16. Higgenbottom and Clay, "Origins."
17. Schibler, Elsner, and Schlumbaum, "Incorporation of Aurochs."
18. Miller and Wettetorm, "Beginnings of Agriculture."
19. Bogaard et al., "Crop Manuring."
20. Schroeder et al., "Unraveling Ancestry."
21. IPBES, *Global Assessment Report.*
22. American Farm Bureau, "Fast Facts."
23. Leahy, "Insect 'Apocalypse.'"
24. Cedar Creek LTER Ecosystem Science Reserve, "E120."
25. Rosenberg et al., "Decline."
26. Berry, *Unsettling of America.*
27. Ingham, Moldenke, and Edwards, *Soil Biology Primer.*
28. Vermeulen, Campbell, and Ingram, "Climate Change."
29. Hawken, *Drawdown.*
30. Montgomery, *Growing a Revolution.*
31. Pywell et al., "Wildlife-Friendly Farming."
32. Gabriel et al., "Scale Matters."
33. "Wes Jackson."
34. Morris, "Land Institute."
35. Rackham, "Hedges."
36. Baudry, Bunce, and Burel, "Hedgerows."
37. McClintock, "James Saules."
38. "Natural Ecosystem Demonstrates Sustainability."
39. Hawken, *Drawdown.*
40. Sortie, Reimer, and Jones, *Oregon Agriculture.*
41. Hine, Pretty, and Twarog, *Organic Agriculture.*
42. United Nations Environment Program, *Mainstreaming Biodiversity.*

CHAPTER 4. FROM PICTURESQUE TO ECOLOGICAL PRESERVATION

1. Bradford, *Of Plymouth Plantation.*
2. Pearsall and Pennington, *Lake District.*
3. Shaw, "Wordsworth and the Sublime."
4. Emerson, *Nature.*
5. Evans, "Soil Erosion."
6. Evans et al., "To Graze."
7. McClelland, *Presenting Nature.*
8. Spence, *Dispossessing the Wilderness.*
9. Ramblers Association, "Access to Open Countryside."

10. Forsthoefel, *Walking to Listen.*
11. Davis, "Psychological Benefits."
12. Sheikh, "How Much Nature Is Enough?"
13. Fleck, *Henry Thoreau and John Muir.*
14. Muir, "American Forests."
15. Nash, "John Muir."
16. "Sierra Club."
17. Simmons, "Crisis in Our National Parks."
18. See Y2Y.net.
19. Mills, "People of Color."
20. Stegner, Wilderness Letter.
21. "NPS Organic Act."
22. Young, *In the Absence of Predators.*
23. Shilling, "Aldo Leopold Listens."
24. Leopold, *Sand County Almanac.*
25. Price, *Introduction to Grand Canyon Geology.*
26. Fox, "What Sparked."
27. NASA, "Hubble Reveals."
28. Ohtomo et al., "Evidence for Biogenic Graphite."
29. Schirrmeister, Gugger, and Donoghue, "Cyanobacteria."
30. Watts, *Cloud-Hidden.*

CHAPTER 5. DOWN THE OREGON COAST TRAIL

1. Walth, *Fire at Eden's Gate.*
2. Lowenthal, *George Perkins Marsh.*
3. Chamberlin, *On the Trail.*
4. Dorman, *Word for Nature.*
5. Thoreau, *Journal;* Thoreau, *Walden,* 25.
6. Fleck, *Henry Thoreau and John Muir.*
7. "Thoreau's Views on Indians"; Thoreau, *Indians of Thoreau.*
8. Iljima and Shaw, "Great 19th Century Timber Heist."
9. "Columbia Flood River Basalts."
10. Wells et al., "Geologic History."
11. Keller, "Deccan Volcanism."
12. Witter et al., *Tracking Prehistoric Cascadia Tsunami Deposits.*
13. Atwater, *Orphan Tsunami.*
14. Gary C. Johnson, email to Donna Sinclair, January 18, 2012.http://
publichistorypdx.org/projects/chinook/lower-columbia-chinook-historical
-timeline/#_ftn2.
15. Beston, *Outermost House.*
16. Carson, *Sea Around Us.*
17. Souder, *On a Farther Shore.*
18. Rodriguez del Rey, Granek, and Sylvester, "Occurrence and Concentration."
19. Foy, "Effects of the Pharmaceutical Industry."
20. Estes and Carswell, "Costs and Benefits."

21. Elakha Alliance, www.elakhaalliance.org.

22. NOAA, "Snapping Shrimp."

23. Pees, "Historic Prices."

24. Hilt, "Incentives in Corporations."

25. Muir, "American Forests."

26. Bullinger, "Yosemite Finally Reckons."

27. Sibthorp et al., "Long-Term Impacts."

28. Kellert and Wilson, *Biophilia Hypothesis.*

29. Goebel, Waters, and O'Rourke, "Late Pleistocene Dispersal."

30. Braje et al., "Fladmark + 40."

31. Curteman, "Geoarchaeological Investigations."

32. Bourgeois and Dott, "Stratigraphy and Sedimentology."

33. Beaulieu and Hughes, *Land Use Geology.*

34. Williams, *References on the American Indian Use of Fire.*

35. Anderson, *Tending the Wild.*

36. Kimmerer, *Braiding Sweetgrass.*

CHAPTER 6. RIVERS OF THE WEST

1. Reimann, "There Used to Be Salmon."

2. Calloway, *One Vast Winter Count.*

3. Meengs and Lackey, "Estimating the Size."

4. Weitkamp, "Hatchery and Wild Salmon."

5. U.S. Geologic Survey, "Moving Mountains."

6. Hogan, *Dwellings.*

7. Engle, Skalicky, and Poirier, *Translocation of Lower Columbia River.*

8. Engle, Skalicky, and Poirier, *Translocation of Lower Columbia River.*

9. Garcia-Navarro, "Tribe Gives Personhood."

10. Horangic, Berry, and Wall, "Influences on Stakeholder Participation."

11. Hawley, *Recovering a Lost River.*

12. Connolly, "Comment Now: Lower Snake River Dams Draft Report."

13. Milly and Dunne, "Colorado River Flow Dwindles."

14. Wegner, "Restoring Glen Canyon."

15. Helfield and Naiman, "Effects of Salmon-Derived Nitrogen."

16. Fleming and Gross, "Breeding Success."

17. Daehnke, *Chinook Resilience.*

18. Gritzner, "Native-American Camas Production."

19. Phillips, *Ethnobotany.*

20. McLaughlin, "Traditions and Diabetes Prevention."

21. Kuhnlein et al., "Arctic Indigenous Peoples"; and Tvesko, "Social Identity and Cultural Change."

22. Hogan, *Radiant Life of Animals.*

23. Deloria, *World We Used to Live In.*

24. Lewis, "Four Deaths."

25. Gahr, "Ethnobiology."

26. Daehnke, *Chinook Resilience.*

27. Gold, "Long Narrows."
28. Aguilar, *When the River Ran Wild!*
29. Bergmann, "Landscapes' Lessons."
30. Aguilar, "Celilo Lives on Paper."
31. "Middle Columbia Fish Advisory."

CHAPTER 7. WILDLIFE

1. IUCN, "Species Extinction."
2. Whitlow, Horne, and Jones, "Basis of Acoustic Discrimination."
3. Ford and Ellis, "Selective Foraging."
4. Wright et al., "Kin-Directed Food Sharing."
5. Simila and Ugarte, "Surface and Underwater Observations."
6. Teixidor, "Cultural Traditions in Orcas."
7. Hildering, "Rub Me Right."
8. U.S. National Marine Fisheries Service, *Recovery Plan.*
9. Fischer, *Human Communication as Narration.*
10. Valdivia, Wolf, and Suckling, "Marine Mammals and Sea Turtles."
11. Hawley, *Recovering a Lost River.*
12. Nestor, *Deep.*
13. Monaco Explorations, "Darewin Project."
14. Welch, "Groundbreaking Effort."
15. Klatt et al., "Bee Pollination."
16. Xerces Society, "Bumble Bee Conservation."
17. Wood, "How Smart Is a Bumblebee?"
18. Tan, "Bees Awe Scientists."
19. Hopwood et al., *How Neonicotinoids Can Kill Bees.*
20. Clevenger, *Trans-Canada Highway.*
21. Green, "Do Urban Growth Boundaries Work?"
22. Christensen, "UGB 101."
23. Rosenberg et al., "Decline."
24. Dresner and Moldenke, "Gardening for Wildlife."
25. Connor, Yoder, and May, "Density-Related Predation."
26. "Black Capped Chickadee."
27. Jones, "Small Bird, Big Mouth."
28. Tremblay and St. Clair, "Permeability of a Heterogeneous Urban Landscape."
29. Clayton, "Domesticated Nature."
30. NatureServe.org.
31. Abram, *Spell of the Sensuous.*
32. Bonnevie et al., "Grid Cells Require."
33. Thompson, "On Wayfinding."

CHAPTER 8. THE HUMAN HEART

1. Ernst and Theimer, "Evaluating the Effects."
2. Chawla, "Life Paths."
3. Wells and Lekies, "Nature and the Life Course."
4. Chawla, "Childhood Nature Connection."

5. Lopez, *Horizon.*

6. Flitcroft et al., "Emergence."

7. Wilson, *Social Conquest of the Earth.*

8. Chenoweth and Stephan, *Why Civil Resistance Works.*

9. Greenpeace, "Koch Industries"; and Goldmacher, "How David Koch."

10. Klein, *This Changes Everything.*

11. Merchant, *Death of Nature;* and Merchant, *Reinventing Eden.*

12. Jackson, *Prosperity Without Growth.*

13. Gatziolis et al., "Elemental Atmospheric Pollution."

14. Schick, "7 Things."

15. Pickett and Wilkinson, *The Spirit Level.*

16. Mikkelson, Gonzalez, and Peterson, "Economic Inequality."

17. Robeyns, "Having Too Much."

18. Hawkin, *Drawdown.*

19. Argyle, *Psychology of Happiness.*

20. Merkel, *Radical Simplicity.*

21. Peat, *Gentle Action.*

22. Nicol, *Subtle Activism.*

23. Peng et al., "Individualist-Collectivist Differences."

24. Stone, "Should Trees Have Standing."

25. See www.centerforenvironmentalrights.org/rights-of-nature-law-library.

26. Bland, "Fishermen Sue."

27. Albert, "Te Awa Tupua."

28. White, "Historical Roots."

29. Boyd and Nishimura, "Shinto Perspectives."

30. Rots, "Sacred Forests, Sacred Nation."

31. Rappaport, *Ritual and Religion.*

32. Fréger, *Wilder Mann.*

33. Lappé, "Shifting the Frame."

34. Macy and Johnstone, "Active Hope."

Bibliography

Abram, David. *The Spell of the Sensuous: Perception and Language in a More-Than-Human World.* New York: Vintage Books, 1996.

Agee, James. "Issues and Impacts of Redwood National Park Expansion." *Environmental Management* 4, no. 5 (1980): 407–23.

Aguilar, George W., Sr. "Celilo Lives on Paper." *Oregon Historical Quarterly* 108, no. 4 (2007), Remembering Celilo Falls: 606–13.

Aguilar, George W., Sr. *When the River Ran Wild! Indian Traditions on the Mid-Columbia and the Warm Springs Reservation.* Oregon Historical Society; in association with University of Washington Press, Portland, Seattle, 2005.

Albert, Gerrard. "Te Awa Tupua: The Journey to Return the Whanganui River to the Whanganui River." 20th International River Symposium and Environmental Flows Conference, Brisbane, Australia, 2017.

American Farm Bureau. "Fast Facts About Agriculture and Food." www.fb.org/newsroom/fast-facts.

Anderson, Kat. *Tending the Wild: Native American Knowledge and the Management of California's Natural Resources.* Berkeley: University of California Press, 2005.

Argyle, Michael. *The Psychology of Happiness.* Center for the New American Dream, 1987.

Atwater, Brian F., Musumi-Rokkaku Satoko, Satake Kenji, Udea Kazue, and David K. Yamaguichi. *The Orphan Tsunami of 1700: Japanese Clues to a Parent Earthquake in North America.* U.S. Geological Survey; in association with University of Washington Press, Reston, VA, Seattle, 2005.

Au, Whitlow, W. L., John K. Horne, and Christopher Jones. "Basis of Acoustic Discrimination of Chinook Salmon from Other Salmons by Ocholocating *Orcinus orca.*" *Journal of the Acoustical Society of America* 128, no. 4 (2010): 2225–32.

Bakeless, John. *America as Seen by Its First Explorers: The Eyes of Discovery.* Mineola, NY: Dover, 2011.

Barnes, Simon. *Rewild Yourself: Making Nature More Visible in Our Lives.* New York: Pegasus Books, 2019.

Baudry, Jacques, R. G. H. Bunce, and Françoise Burel. "Hedgerows: An International Perspective on Their Origin, Function and Management." *Journal of Environmental Management* 60, no. 1 (2000): 7–22.

Beaulieu, John D., and Paul W. Hughes. *Land Use Geology of Western Curry County, Oregon.* Bulletin 90. State of Oregon, Department of Geology and Mineral Industries, 1976.

Bergmann, Mathias D. "Landscapes' Lessons: Native American Cultural Geography in Nineteenth-Century Oregon and Washington." *Other Ways of Knowing* 2 (2016): 45–85.

Berry, Wendell. *The Unsettling of America*. Berkeley, CA: Counterpoint Press, 1977.

Beston, Henry. *The Outermost House*. Garden City, NY: Doubleday, Doran, 1928.

Binkley, Daniel, Tom Sisk, Carol Chambers, Judy Springer, and William Block. "The Role of Old-Growth Forests in Frequent-Fire Landscapes." *Ecology and Society* 12, no. 2 (2007): art. 18.

"Black-Capped Chickadee." All About Birds, Cornell Lab. https://www .allaboutbirds.org/guide/Black-capped_Chickadee/overview.

Bland, Alastair. "Fishermen Sue Big Oil for Its Role in Climate Change." *The Salt* (blog). *NPR*, December 4, 2018.

Bogaard, Amy, Rebecca Fraser, Tim H. E. Heaton, Michael Wallace, Petra Vaiglova, Michael Charles, Glynis Jones et al. "Crop Manuring and Intensive Land Management by Europe's First Farmers." *Proceedings of the National Academy of Sciences* 110, no. 31 (2013): 12589–94.

Bonnevie, Tora, Benjamin Dunn, Marianne Fyhn, Torkel Hafting, Dori Derdikman, John Kubic, Yasser Roudi et al. "Grid Cells Require Excitatory Drive from the Hippocampus." *Nature Neuroscience* 16 (2013): 309–17.

Bourgeois, Joanne, and R. H. Dott Jr. "Stratigraphy and Sedimentology of Upper Cretaceous Rocks in Coastal Southwest Oregon: Evidence for Wrench-Fault Tectonics in a Postulated Accretionary Terrane." *Geological Society of America Bulletin* 96, no. 8 (1985): 1007–19.

Boyd, James, and Tetsuya Nishimura. "Shinto Perspectives in Miyazaki's Anime Film *Spirited Away*." *Journal of Religion and Film* 8, no. 3 (2004): art. 4.

Bradford, William. *Of Plymouth Plantation, 1620–1647*. 1856. New York: Knopf, 1963.

Braje, Todd J., Jon Erlandson, Torben Rick, and Loren G. Davis. "Fladmark + 40: What Have We Learned About a Potential Pacific Coast Peopling of the Americas?" *American Antiquity* 85, no. 1 (2019): 1–21.

Broome, Richard. *Aboriginal Australians*. 2nd ed. Australia: Allen and Unwin, 1994.

Bullinger, Jake. "Yosemite Finally Reckons with Its Discriminatory Past." *Outside*, August 23, 2018.

Bürgi, Matthias, Lars Östlund, and David J. Mladenoff. "Legacy Effects of Human Land Use: Ecosystems as Time-Lagged Systems." *Ecosystems* 20, no. 1 (2017): 94–103.

Calloway, Colin. *One Vast Winter Count: The Native American West Before Lewis and Clark*. Lincoln: University of Nebraska Press, 2003.

Carson, Rachel. *The Sea Around Us*. New York: Oxford University Press, 1950.

Cedar Creek LTER Ecosystem Science Reserve. "E120—Biodiversity II: Effects of Plant Biodiversity on Population and Ecosystem Processes." University of Minnesota, 2023.

Chamberlin, Silas. *On the Trail: A History of American Hiking*. New Haven, CT: Yale University Press, 2016.

Chauvet, Jean-Marie, Eliette Brunel Deschamps, and Christian Hilaire. *La Grotte Chauveta Vallon-Pont-D'Arc*. Paris: Seuil, 1996.

Chawla, Louise. "Childhood Nature Connection and Constructive Hope: A

Review of Research on Connecting with Nature and Coping with Environmental Loss." *People and Nature* 2, no. 3 (2020): 619–42.

Chawla, Louise. "Life Paths into Effective Environmental Action." *Journal of Environmental Education* 31, no. 1 (1999): 15–26.

Chenoweth, Erica, and Maria J. Stephan. *Why Civil Resistance Works: The Strategic Logic of Nonviolent Conflict.* New York: Columbia University Press, 2011.

Christensen, Nick. "UGB 101: Everything You Wanted to Know About the Urban Growth Boundary, but Were Afraid to Ask." *Metro News,* January 9, 2018.

Clayton, Susan. "Domesticated Nature: Motivations for Gardening and Perceptions of Environmental Impact." *Journal of Environmental Psychology* 27, no. 3 (2007): 215–24.

Clevenger, A. P., D. Duke, R. Haddock, and R. Ament. *Trans-Canada Highway Wildlife and Monitoring Research, Annual Report 2012–13.* Prepared for Parks Canada Agency, Radium Hot Springs, BC, 2013.

Clottes, Jean. *What Is Paleolithic Art? Cave Paintings and the Dawn of Human Creativity.* Chicago: University of Chicago Press, 2016.

Clottes, Jean, and J. David Lewis-Williams. *The Shamans of Prehistory: Trance and Magic in the Painted Caves.* New York: Harry N. Abrams, 1998.

"Columbia Flood River Basalts." Volcano World. Oregon State University, 2010. https://volcano.oregonstate.edu/volcano.

Connolly, Chris. "Comment Now: Lower Snake River Dams Draft Report." Northwest Energy Coalition. https://nwenergy.org/featured/comment-now-lower -snake-river-dams-draft-report/.

Connor, Edward F., James M. Yoder, and Julie A. May. "Density-Related Predation by the Carolina Chickadee, *Poecile carolinensis,* on the Leaf-Mining Moth, *Cameraria hamadryadella,* at Three Spatial Scales." *Oikos* 87, no. 1 (1999): 105–12.

Cortazar, Ryan Z. "Scholar: Cave Paintings Show Religious Sophistication." *Harvard Gazette,* April 26, 2007. https://news.harvard.edu/gazette/story/2007/04/ scholar-cave-paintings-show-religious-sophistication/.

Cosmides, Leda, and Jon Tooby. "Evolutionary Psychology: A Primer." Center for Evolutionary Psychology, UC Santa Barbara, 1997.

Crotteau, Justin S., and Christopher R. Keyes. "Restoration Treatments Improve Overstory Tree Resistance Attributes and Growth in a Ponderosa Pine/Douglas-Fir Forest." *Forests* 11, no. 5 (2020): art. 574.

Cummings, Vicky, Magdalena S. Midgley, and Chris Scarre. "Chambered Tombs and Passage Graves of Western and Northern Europe." In *Oxford Handbook of Neolithic Europe,* edited by Chris Fowler, Jan Harding, and Daniela Hoffman, 813–38. Oxford: Oxford University Press, 2015.

Curry, Andrew. "The First Europeans Weren't Who You Might Think." *National Geographic,* August 2019.

Curteman, Jessica Anne. "Geoarchaeological Investigations at the Devils Kitchen Site (35CS9), Southern Oregon Coast." MA thesis, Oregon State University, 2016.

Daehnke, Jon. *Chinook Resilience: Heritage and Cultural Revitalization on the Lower Columbia River.* Seattle: University of Washington Press, 2017.

Davis, John. "Psychological Benefits of Nature Experiences: An Outline of Research and Theory." Naropa University and School of Lost Borders, July 2004.

Delage, Christophe. "Comments on a Recent Challenge of the Authenticity of the La Marche Engravings." *Fornvännen: Journal of Swedish Antiquarian Research* 111, no. 3 (2016): 192–97.

Deloria, Vine, Jr. *The World We Used to Live In: Remembering the Powers of the Medicine Men.* Golden, CO: Fulcrum, 2006.

D'Errico, Francesco, and Christopher S. Henshilwood. "The Origin of Symbolically Mediated Behavior: From Antagonistic Scenarios to a Unified Research Strategy." In *Homo Symbolicus: The Dawn of Language, Imagination, and Spirituality,* edited by Christopher S. Henshilwood and Francesco d'Errico, 49–73. Amsterdam: Benjamins, 2011.

D'Huy, Julien, and Jean-Loïc Le Quellec. "Les animaux 'fléchés' à Lascaux: Nouvelle proposition d'interprétation." *Préhistoire du Sud-Ouest* 18, no. 2 (2010): 161–70.

Dorman, Robert L. *A Word for Nature: Four Pioneering Environmental Advocates, 1845–1913.* Chapel Hill: University of North Carolina Press, 1998.

Dresner, Marion, and Andrew Moldenke. "Gardening for Wildlife: Tree Canopy and Small-Scale Planting Influences on Arthropod and Bird Abundance." *Cities and the Environment* 10, no. 1 (2017): art. 9.

Dupouey, Jean-Luc, Etienne Dambrine, Jean-Denis Laffite, and C. Moares. "Irreversible Impact of Past Land Use on Forest Soils and Biodiversity." *Ecology* 83, no. 11 (2002): 2978–84.

Egan, Timothy. *The Big Burn: Teddy Roosevelt and the Fire That Saved America.* Boston: Mariner Books, 2010.

Emerson, Ralph Waldo. *Nature.* Boston: James Munroe and Company, 1836.

Engle, Rod, Joseph Skalicky, and Jen Poirier. *Translocation of Lower Columbia River Fall Chinook Salmon* (Oncorhynchus tshawytscha) *in the Year of Condit Dam Removal and Year One Post-Removal Assessment.* U.S. Fish and Wildlife Service, Columbia River Fisheries Program Office, Vancouver, WA, 2013.

Ernst, Julie, and Stefan Theimer. "Evaluating the Effects of Environmental Education Programming on Connectedness to Nature." *Environmental Education Research* 17, no. 5 (2011): 577–98.

Estes, James A., and Lilian P. Carswell. "Costs and Benefits of Living with Predators." *Science* 368, no. 6496 (2020): 1178–80.

Evans, Darren M., Stephen M. Redpath, David A. Elston, Sharon A. Evans, Ruth J. Mitchell, and Peter Dennis. "To Graze or Not to Graze? Sheep, Voles, Forestry, and Nature Conservation in the British Uplands." *Journal of Applied Ecology* 43, no. 3 (2006): 499–505.

Evans, Robert. "Soil Erosion in the UK Initiated by Grazing Animals: A Need for a National Survey." *Applied Geography* 17, no. 2 (1997): 127–41.

Ewing, Holly A., Kathleen C. Weathers, Pamela H. Templer, Todd E. Dawson, Mary K. Firestone, Amanda M. Elliott, and Vanessa K. S. Boukili. "Fog Water and Ecosystem Function." *Ecosystems* 12 (2009): 417–33.

Fisher, Walter R. *Human Communication as Narration: Toward a Philosophy of Reason, Value, and Action*. Columbia: University of South Carolina Press, 1989.

Flannery, Kent, and Joyce Marcus. *The Creation of Inequality: How Our Prehistoric Ancestors Set the Stage for Monarchy, Slavery, and Empire*. Cambridge, MA: Harvard University Press, 2012.

Flannery, Tim. *The Future Eaters: An Ecological History of the Australasian Lands and People*. Sydney: New Holland, 1997.

Flavelle, Christopher, and Henry Fountain. "Oregon Burns in Places That Normally Don't See Fire." *New York Times,* September 13, 2020.

Fleck, Richard F. *Henry Thoreau and John Muir Among the Indians*. Hamden, CT: Archon Books, 1985.

Fleming, Ian A., and Mart R. Gross. "Breeding Success of Hatchery and Wild Coho Salmon (*Oncorhynchus kisutch*) in Competition." *Ecological Applications* 3, no. 2 (1993): 230–45.

Fleur, Nicholas St. "Earliest Known Human Footprints in North America Found on Canadian Island." *New York Times,* March 28, 2018.

Flitcroft, Rebecca L., Lee K. Cerveny, Bernard T. Bormann, Jane E. Smith, Stanley T. Asah, and A. Paige Fischer. "The Emergence of Watershed and Forest Collaboratives." In *People, Forests, and Change,* edited by Deanna H. Olson and Beatrice Van Horne, 116–30. Washington, DC: Island Press/Center for Resource Economics, 2017.

Ford, John K. B., and Graeme M. Ellis. "Selective Foraging by Fish-Eating Killer Whales *Orcinus orca* in British Columbia." *Marine Ecology Progress Series* 316 (2006): 185–99.

Forsthoefel, Andrew. *Walking to Listen: 4,000 Miles Across America, One Story at a Time*. New York: Bloomsbury, 2017.

Fox, Douglas. "What Sparked the Cambrian Explosion?" *Nature* 530 (2016): 268–70.

Foy, Anne. "The Effects of the Pharmaceutical Industry on the Ocean." *Ocean Crusaders,* October 28, 2015. https://oceancrusaders.org/pharmaceutical-ocean/.

Fréger, Charles. *Wilder Mann; ou la figure du sauvage*. London: Thames & Hudson, 2012.

Friedman, Steven K., and Peter B. Reich. "Regional Legacies of Logging: Departure from Presettlement Forest Conditions in Northern Minnesota." *Ecological Applications* 15, no. 2 (2005): 726–44.

Fuentas, Oscar. "The Depiction of the Individual in Prehistory: Human Representations in Magdalenian Societies." *Antiquity* 87, no. 338 (2013): 985–1000.

Gabriel, Doreen, Steven M. Sait, Jenny A. Hodgson, Ulrich Schmutz, William E. Kunin, and Tim G. Benton. "Scale Matters: The Impact of Organic Farming on Biodiversity at Different Spatial Scales." *Ecology Letters* 13, no. 7 (2010): 858–69.

Gahr, D. Ann Trieu. "Ethnobiology: Nonfishing Subsistence and Production." In *Chinookan People of Lower Columbia,* edited by Robert T. Boyd, Kenneth M. Ames, and Tony A. Johnson, 63–79. Seattle: University of Washington Press.

Garcia-Navarro, Lulu. "Tribe Gives Personhood to Klamath River." *NPR, Weekend Edition Sunday,* September 29, 2019.

Gatziolis, Demetrios, Sarah Jovan, Geoffrey Donovan, Michael Amacher, and Vicente Monleon. "Elemental Atmospheric Pollution Assessment via Moss-based Measurements in Portland, Oregon." Gen. Tech. Rep. PNW-GTR-935. Portland, OR: U.S. Department of Agriculture, Forest Service, Pacific Northwest Research Station, 2016.

George, Alison. "Code Hidden in Stone Age Art May Be the Root of Human Writing." *New Scientist,* November 9, 2016.

"Giant Logged Long Ago but Not Forgotten." *Seattle Times,* September 4, 2011.

Goebel, Ted, Michael R. Waters, and Dennis H. O'Rourke. "The Late Pleistocene Dispersal of Modern Humans in the Americas." *Science* 319, no. 5869 (2008): 1497–502.

Gold, Pat C. "The Long Narrows: The Forgotten Geographic and Cultural Wonder." *Oregon Historical Quarterly* 108, no. 4 (2007), Remembering Celilo Falls.

Goldmacher, Shane. "How David Koch and His Brother Shaped American Politics." *New York Times,* August 23, 2019.

Gray, Andrew N., and Thomas R. Whittier. "Carbon Stocks and Changes on Pacific Northwest National Forests and the Role of Disturbance, Management, and Growth." *Forest Ecology and Management* 328 (2014): 167–78.

Green, Jared L. "Do Urban Growth Boundaries Work to Prevent Sprawl?" *The Dirt: Uniting the Built and Natural Environments* (blog). ASLA, 2015. https://www.smartcitiesdive.com/ex/sustainablecitiescollective/do-urban-growth-boundaries-work/1070356/.

Greenpeace. "Koch Industries: Still Funding Climate Denial, 2011 Update." https://www.greenpeace.org/usa/wp-content/uploads/2015/07/Koch-Ind-Still-Fueling-Climate-Denial.pdf?a1481f.

Gritzner, Janet. "Native-American Camas Production and Trade in the Pacific Northwest and Northern Rocky Mountains." *Journal of Cultural Geography* 14, no. 2 (1994): 33–50.

Groenen, Marc. "Images de mains dans la préhistoire." *La part de l'œil: Revue de pensée des arts plastiques,* L'art et la fonction symbolique, 25–26 (2011): 56–69. https://hal.science/hal-02613420/document.

Grosvenor, Melville Bell. "World's Tallest Tree Discovered." *National Geographic* 126, no. 1 (July 1964): 1–9.

Harari, Yuval Noah. *Sapiens: A Brief History of Humankind.* New York: McClelland & Stewart, 2014.

Hart, Melissa A., and David J. Sailor. "Quantifying the Influence of Land-Use and Surface Characteristics on Spatial Variability in the Urban Heat Island." *Theoretical and Applied Climatology* 95 (2008): 397–406.

Hassett, Brenna. *Built on Bones: 15,000 Years of Urban Life and Death.* London: Bloomsbury Sigma, 2017.

Hawken, Paul, ed. *Drawdown: The Most Comprehensive Plan Ever Proposed to Reverse Global Warming.* New York: Penguin, 2017.

Hawley, Steven. *Recovering a Lost River: Removing Dams, Rewilding Salmon, Revitalizing Communities*. Boston: Beacon Press, 2012.

Helfield, James M., and Robert J. Naiman. "Effects of Salmon-Derived Nitrogen on Riparian Forest Growth and Implications for Stream Productivity." *Ecology* 82, no. 9 (2001): 2403–9.

Higgenbottom, Gal, and Roger Clay. "Origins of Standing Stone Astronomy in Britain: New Quantitative Techniques for the Study of Archaeoastronomy." *Journal of Archaeological Science* 9 (October 2016): 249–58.

Hildering, Jackie. "Rub Me Right: 'Beach-Rubbing' Behaviour of Northern Resident Orca." *Marine Detective* (blog), February 8, 2015. https://themarinedetective.com/2015/02/08/rub-me-right-beach-rubbing-behaviour-of-northern-resident-orca/.

Hilt, Eric. "Incentives in Corporations: Evidence from the American Whaling Industry." *Journal of Law and Economics* 49, no. 1 (April 2006):197–227.

Hine, Rachel, Jules Pretty, and Sophia Twarog. *Organic Agriculture and Food Security in Africa*. UNEP-UNCTAD Capacity-Building Task Force on Trade, Environment and Development, 2008.

Hogan, Linda. *Dwellings: A Spiritual History of the Living World*. New York: W. W. Norton, 1995.

Hogan, Linda. *The Radiant Life of Animals*. Boston: Beacon Press, 2020.

Hopwood, Jennifer, Aimee Code, Mace Vaughan, David Biddinger, Matthew Shepherd, Scott Hoffman Black, Eric Lee-Mäder, and Celeste Mazzacano. *How Neonicotinoids Can Kill Bees*. Xerces Society, 2016.

Horangic, Alexandra, Kate A. Berry, and Tamara Wall. "Influences on Stakeholder Participation in Water Negotiations: A Case Study from the Klamath Basin." *Society and Natural Resources* 29, no. 12 (2016): 1421–35.

Hudson, Dale A. "Sierra Club v. Department of Interior: The Fight to Preserve the Redwood National Park." *Ecology Law Quarterly* 7, no. 3 (1979): 781–859.

Iljima, T. J., and Jane S. Shaw. "The Great 19th Century Timber Heist Revisited." *The Freeman* 40, no. 4 (April 1990): 137–41.

Ingham, Elaine R., Andrew R. Moldenke, and Clive A. Edwards. *Soil Biology Primer*. Ankeny, IA: Soil and Water Conservation Society, 2000.

IPBES. *Global Assessment Report on Biodiversity and Ecosystem Services of the Intergovernmental Science-Policy Platform on Biodiversity and Ecosystem Services,* edited by Eduardo S. Brondizio, Josef Settele, Sandra Díaz, and Hien T. Ngo. IPBES Secretariat, Bonn, Germany, 2019. https://doi.org/10.5281/zenodo.3831673.

Isbell, Forest Isaac, David Tilman, Peter B. Reich, and Adam Thomas Clark. "Deficits of Biodiversity and Productivity Linger a Century After Agricultural Abandonment." *Nature, Ecology, and Evolution* 3 (2019): 1533–38.

IUCN. "Species Extinction." IUCN. https://www.iucn.org/content/species-extinction.

Jackson, Tim. *Prosperity Without Growth? A Transition to a Sustainable Economy*. London: Sustainable Development Commission, 2009.

Jones, Harrison. "Small Bird, Big Mouth: The Surprisingly Complex Language of the Carolina Chickadee." University of Florida IFAS, 2015.

Kahn, Razib. "Foragers to Farmers: A Tale of Collective Action?" *Discover,* March 18, 2011.

Keller, Gerta. "Deccan Volcanism Caused the Mass Extinction 66 Million Years Ago." Department of Geosciences, Princeton University, September 23, 2020.

Kellert, Stephen, and E. O. Wilson, eds. *The Biophilia Hypothesis.* Washington, DC: Island Press, 1995.

Kimmerer, Robin Wall. *Braiding Sweetgrass: Indigenous Wisdom, Scientific Knowledge, and the Teachings of Plants.* Minneapolis, MN: Milkweed, 2013.

Klatt, Björn K., Andrea Holzschuh, Catrin Westphal, Yann Clough, Inga Smit, Elka Pawelzik, and Teja Tacharntke. "Bee Pollination Improves Crop Quality, Shelf Life, and Commercial Value." *Proceedings of the Royal Society B: Biological Sciences* 281 (January 22, 2014). https://doi.org/10.1098/rspb.2013.2440.

Klein, Naomi. *This Changes Everything: Capitalism vs. the Climate.* New York: Simon & Schuster, 2014.

Kuhnlein, Harriet V., Olivier Receveur, Rula Soueida, and Grace M. Egeland. "Arctic Indigenous Peoples Experience the Nutrition Transition with Changing Dietary Patterns and Obesity." *Journal of Nutrition* 134, no. 6 (2004): 1447–53.

Lappé, Frances Moore. "Shifting the Frame to Imagine a Different World: An Interview with Frances Moore Lappé." By Lorelei Hanson and Patricia Ballamingi. *Aurora,* October 1, 2010.

Law, Beverly E., Tara W. Hudiburg, Logan T. Berner, Jeffrey J. Kent, Polly C. Buotte, and Mark E. Harmon. "Land Use Strategies to Mitigate Climate Change in Carbon Dense Temperate Forests." *Proceedings of the National Academy of Sciences* 115, no. 14 (2018): 3663–68.

Leahy, Stephen. "Insect 'Apocalypse' in U.S. Driven by 50x Increase in Toxic Pesticides." *National Geographic,* August 6, 2019.

Leopold, Aldo. *A Sand County Almanac.* New York: Oxford University Press, 1949.

Lewis, David. "Cascadia Cave." *Oregon Encyclopedia.* September 16, 2022. www.oregonencyclopedia.org/articles/cascadia_cave/#author-154-info.

Lewis, David. "Four Deaths: The Near Destruction of Western Oregon Tribes and Native Lifeways, Removal to the Reservation, and Erasure from History." *Oregon Historical Quarterly* 115, no. 3 (2014): 414–37.

Lewis-Williams, J. David. *Conceiving God: The Cognitive Origin and Evolution of Religion.* London: Thames & Hudson, 2010.

Lewis-Williams, J. David. *The Mind in the Cave.* London: Thames & Hudson, 2002.

Lewis-Williams, J. David, and David Pierce. *Inside the Neolithic Mind: Consciousness, Cosmos, and the Realm of the Gods.* London: Thames & Hudson, 2005.

Lipson, Mark, Anna Szecsenyi-Nagy, and David Reich. "Parallel Paleogenomic Transects Reveal Complex Genetic History of Early European Farmers." *Nature* 551 (November 16, 2017): 368–72.

Lopez, Barry. *Horizon.* New York: Knopf, 2019.

Lowenthal, David. *George Perkins Marsh: Prophet of Conservation.* Seattle: University of Washington Press, 2000.

Lutz, James A., Tucker J. Furniss, Daniel J. Johnson, Stuart J. Davies, David Allen, Alfonso Alonso, Kristina J. Anderson-Teixeira et al. "Global Importance of Large-Diameter Trees." *Global Ecology and Biogeography* 27, no. 7 (2018): 849–64.

Macy, Joanna, and Chris Johnstone. *Active Hope: How to Face the Mess We're in with Unexpected Resilience and Creative Power.* Novato, CA: New World Library, 2022.

Manning, Richard. *Against the Grain: How Agriculture Has Hijacked Civilization.* New York: North Point Press, 2015.

Marshack, Alexander. "The Taï Plaque and Calendrical Notation in the Upper Paleolithic." *Cambridge Archaeological Journal* 1, no. 1 (1991): 25–61.

McClelland, Linda Flint. *Presenting Nature: The Historic Landscape Design of the National Park.* National Park Service, U.S. Department of Interior, 1993.

McClintock, Thomas C. "James Saules, Peter Burnett, and the Oregon Black Exclusion Law of June 1844." *Pacific Northwest Quarterly* 86, no. 3 (1995): 121–30.

McLaughlin, Sue. "Traditions and Diabetes Prevention: A Healthy Path for Native Americans." *Diabetes Spectrum* 23, no. 4 (2010): 272–77.

Meany, Edmond S., ed. "Diary of Wilkes in the Northwest." *Washington Historical Quarterly* 16, nos. 2 and 4 (1925): 140–45, 297–98.

Meengs, Chad C., and Robert T. Lackey. "Estimating the Size of Historical Oregon Salmon Runs." *Reviews in Fisheries Science* 13, no. 1 (2005): 51–66.

Meigs, Garrett, Christopher Dunn, Sean Parks, and Meg Krawchuk. "Influence of Topography and Fuels on Fire Refugia Probability Under Varying Fire Weather Conditions in Forests of the Pacific Northwest, USA." *Canadian Journal of Forestry Research* 50, no. 7 (2020): 636–47.

Merchant, Carolyn. *The Death of Nature: Women, Ecology, and the Scientific Revolution.* New York: Harper & Row, 1980.

Merchant, Carolyn. *Reinventing Eden: The Fate of Nature in Western Culture.* New York: Taylor and Francis, 2013.

Merhaut, Donald J. "How Do Large Trees, Such as Redwoods, Get Water from Their Roots to Their Leaves?" *Scientific American,* February 8, 1999.

Merkel, Jim. *Radical Simplicity: Small Footprints on a Finite Earth.* Gabriola Island, BC: New Society Publishers, 2003.

"Middle Columbia Fish Advisory." Fish and Shellfish Consumption. Oregon Health Authority. Oregon.gov.

Mikkelson, Gregory M., Andrew Gonzalez, and Garry D. Peterson. "Economic Inequality Predicts Biodiversity Loss." *PLoS ONE* 2, no. 5 (May 16, 2007).

Miller, Naomi, and Wilma Wettettorm. "The Beginnings of Agriculture: The Ancient Near East and Africa." In *The Cambridge World History of Food,* edited by Kenneth F. Kiple and Kriemhild Conneé Ornelas, 1123–39. Cambridge: Cambridge University Press, 2000.

Mills, James E. "These People of Color Transformed U.S. National Parks." *National Geographic,* August 5, 2020.

Milly, Paul C. D., and Krista A. Dunne. "Colorado River Flow Dwindles as Warming-Driven Loss of Reflective Snow Energizes Evaporation." *Science* 367, no. 6483 (2020): 1252–55.

Minc, Leah, and Kevin Smith. "The Spirit of Survival: Cultural Responses to Resource Variability in North Alaska." In *Bad Year Economics: Cultural Responses to Risk and Uncertainty,* edited by Paul Halstead and John M. O'Shea, 8–39. Cambridge: Cambridge University Press 1989.

Monaco Explorations. "Darewin Project: 'Understanding Whale and Dolphin Click Communication.'" https://www.monacoexplorations.org/en/darewin-project-understanding-whale-and-dolphin-click-communication/.

Montgomery, David R. *Growing a Revolution: Bringing Our Soil Back to Life.* New York: W. W. Norton, 2017.

Morris, Bob. "The Land Institute Is Developing Perennial Crops." *Politics in the Zeros* (blog), December 22, 2018. https://polizeros.com/2018/12/22/land-institute-perennial-crops/.

Morrison, Peter, and Frederick Swanson. *Fire History and Pattern in a Cascade Range Landscape.* Gen. Tech. Rep. PNW-GTR-254. Portland, OR, 1990.

Muir, John. "The American Forests." 1897. Republished in *The Atlantic* 80, no. 478 (November 2012): 145–57.

NASA. "Hubble Reveals Observable Universe Contains 10 Times More Galaxies Than Previously Thought." October 13, 2016. https://www.nasa.gov/feature/goddard/2016/hubble-reveals-observable-universe-contains-10-times-more-galaxies-than-previously-thought.

Nash, Roderick. "John Muir: Publicizer." Chap. 8 in *Wilderness and the American Mind.* New Haven, CT: Yale University Press, 2014.

"Natural Ecosystem Demonstrates Sustainability." *AskNature,* June 3, 2020. https://asknature.org/strategy/natural-ecosystem-demonstrates-sustainability/.

Nestor, James. *Deep: Freediving, Renegade Science, and What the Ocean Tells Us About Ourselves.* Boston: Houghton Mifflin Harcourt, 2014.

Nicol, David. *Subtle Activism: The Inner Dimension of Social and Planetary Transformation.* Albany: State University of New York Press, 2015.

NOAA (National Oceanic and Atmospheric Administration). "Snapping Shrimp May Ring 'Dinner Bell' for Gray Whales Off the Oregon Coast." NOAA Research Notes, February 12, 2018. https://www.nationalgeographic.com/animals/article/snapping-shrimp-eastern-pacific-gray-whales-feeding-mysids-oregon-coast-spd.

"NPS Organic Act." Office of Congressional and Legislative Affairs, U.S. Department of the Interior. https://www.doi.gov/ocl/nps-organic-act.

Ohtomo, Yoko, Takeshi Kakegawa, Akizumi Ishida, and Minik T. Rosing. "Evidence for Biogenic Graphite in Early Archaean Isua Metasedimentary Rocks." *Nature Geoscience* 7 (2014): 25–28.

"The Orkney Venus." The Links of Noltland Excavations, Westray. *Orkneyjar,* 2009.

Otterbein, Keith. "The Archaeology of War." *Anthropology News* 44, no. 9 (July 2009).

Pearsall, W. H., and Winifred Pennington. *The Lake District: A Landscape History*. London: Collins, 1973.

Peat, F. David. *Gentle Action: Bringing Creative Change to a Turbulent World*. Pari Grosseto, Italy: Pari Pub., 2008.

Pees, Samuel T. "Historic Prices." *Oil History*. www.petroleumhistory.org/OilHistory/pages/Whale/prices.html.

Peng, Xiang, Zhang Haibo, Geng Liuna, Zhou Kexin, and Wu Yuping. "Individualist-Collectivist Differences in Climate Change Inaction: The Role of Perceived Intractability." *Frontiers in Psychology* 10 (2019): 187.

Phillips, Patricia Whereat. *Ethnobotany of the Coos, Lower Umpqua, and Siuslaw Indians*. Corvallis: Oregon State University Press, 2016.

Price, L. Greer. *An Introduction to Grand Canyon Geology*. Grand Canyon, AZ: Grand Canyon Association, 1999.

Prinsloo, Daniel R. "Mycorrhizal Fungi Shaped the Evolution of Terrestrial Plants." *Plant Science Research Weekly* (blog), November 17, 2017.

Pywell, Richard, Matthew Heard, Ben Woodcock, Shelley Hinsley, Lucy Ridding, Marek Nowakowski, and James M. Bullock. "Wildlife-Friendly Farming Increases Crop Yield: Evidence for Ecological Intensification." *Proceedings of the Royal Society B: Biological Sciences* 282, no. 18616 (2015).

Rackham, Oliver. "Hedges and Hedgerow Trees in Britain: A Thousand Years of Agroforestry." *Social Forestry Network,* network paper 8c, 1989.

Ramblers Association. "Access to Open Countryside." *Ramblers,* August 18, 2020. Accessed April 4, 2023. https://www.ramblers.org.uk/.

Rappaport, Roy A. *Ritual and Religion in the Making of Humanity*. Cambridge: Cambridge University Press, 1999.

Raux, Pascal. "De la grotte ornée à la sacralisation des objets d'art mobiliers." *Bulletin de la S.E.R.P.E.,* no. 58 (2008): 80–91.

Reimann, Matt. "There Used to Be Salmon as Big as Golden Retrievers in the Columbia River, but Dams Killed Them Off." *Timeline,* April 26, 2017. https://timeline.com/there-used-to-be-salmon-as-big-as-golden-retrievers-in-the-columbia-river-but-dams-killed-them-off-20854d1f971e.

Richards, Colin, Andrew Meirion Jones, Ann MacSween, Alison Sheridan, Elaine Dunbar, Paula Reimer, Alex Baylis, Seren Griffiths, and Alasdair Whittle. "Settlement Duration and Materiality: Formal Chronological Models for the Development of Barnhouse, a Grooved Ware Settlement in Orkney." *Proceedings of the Prehistoric Society* 82 (2016): 193–225.

Robeyns, Ingrid. "Having Too Much." *Nomos* 58, Wealth (2017): 1–44.

Rodriguez del Rey, Zoe, Elise Granek, and Steve Sylvester. "Occurrence and Concentration of Caffeine in Oregon Coastal Waters." *Marine Pollution Bulletin* 64, no. 7 (2012): 1417–24.

Rosenberg, Kenneth, Adriaan Dokter, Peter Blancher, John Sauer, Adam Smith, Paul Smith, Jessica Stanton, et al. "Decline of the North American Avifauna." *Science* 366, no. 6461 (2019): 120–24.

Rots, Aike. "Sacred Forests, Sacred Nation: The Shinto Environmentalist Paradigm and the Rediscovery of 'Chinju no Mori.'" *Japanese Journal of Religious Studies* 42, no. 2 (2015): 205–33.

Rutherford, Adam. *A Brief History of Everyone Who Ever Lived: The Human Story Retold Through Our Genes.* New York: The Experiment, 2017.

Sahlins, Marshall. *Stone Age Economics.* New York: De Gruyter, 1972.

Sauter, Disa, Frank Eisner, Paul Ekman, and Sophie Scott. "Cross-Cultural Recognition of Basic Emotions Through Nonverbal Emotional Vocalizations." *Proceedings of the National Academy of Sciences* 107, no. 6 (2010): 2408–12.

Scarre, Chris. *Monuments and Landscapes in Atlantic Europe: Perceptions and Society During the Neolithic and Early Bronze Age.* Milton, UK: Taylor & Francis, 2005.

Schibler, Jörg, Julia Elsner, and Angela Schlumbaum. "Incorporation of Aurochs into a Cattle Herd in Neolithic Europe: Single Event or Breeding?" *Scientific Reports* 4, no. 5798 (2014).

Schick, Tony. "7 Things You Need to Know About Portland's Toxic Air Situation." *Oregon Public Broadcasting,* March 3, 2016.

Schirrmeister, Bettina, Muriel Gugger, and Philip C. J. Donoghue. "Cyanobacteria and the Great Oxidation Event: Evidence from Genes and Fossils." *Palaeontology* 58, no. 5 (2015): 769–85.

Schroeder, Hannes, Ashot Margaryan, Marzena Szmyt, Bertrand Theulot, Piotr Włodarczak, Simon Rasmussen, Shyam Gopalakrishnan et al. "Unraveling Ancestry, Kinship, and Violence in a Late Neolithic Mass Grave." *Proceedings of the National Academy of Sciences* 116, no. 22 (2019):10705–10.

Schulz Paulsson, Bettina. "Radiocarbon Dates and Bayesian Modeling Support Maritime Diffusion Model for Megaliths in Europe." *Proceedings of the National Academy of Sciences* 116, no. 9 (2019): 3460–65.

Sealy, Dan, Suzanne Guerra, and John Amodio. "Redwood National Park Expansion, Woodstock, Earth Day, and the Kent State Massacre." Proceedings of the 2011 George Wright Society Conference on Parks, Protected Areas, and Cultural Sites. www.georgewright.org/1151sealy.pdf.

Shaw, Philip. "Wordsworth and the Sublime." Discovering Literature: Romantics & Victorians. British Library. May 15, 2014.

Sheikh, Knvul. "How Much Nature Is Enough?" *New York Times,* June 13. 2019.

Shilling, Dan. "Aldo Leopold Listens to the Southwest." *Journal of the Southwest* 51, no. 3 (Autumn 2009): 317–50.

Shillito, Lisa-Marie. Helen L. Whelton, John C. Blong, Dennis L. Jenkins, Thomas J. Connolly, and Ian D. Bull. "Pre-Clovis Occupation of the Americas Identified by Human Fecal Biomarkers in Coprolites from Paisley Caves, Oregon." *Science Advances* 6, no. 29 (July 15, 2020).

Sibthorp, Jim, Karen Paisley, Nathan Furman, and John Gookin. "Long-Term Impacts Attributed to Participation in Wilderness Education: Preliminary Findings from NOLS." *Research in Outdoor Education* 9, no. 10 (2008): 86–103.

"Sierra Club Grand Canyon Dam Advertisements, 1966." *Energy History.* Yale

University. https://energyhistory.yale.edu/library-item/sierra-club-grand-canyon-dam-advertisements-1966.

Simard, Suzanne. *Finding the Mother Tree: Discovering the Wisdom of the Forest.* New York: Knopf, 2021.

Simard, Suzanne. "How Trees Talk to Each Other." TED talk. July 22, 2016. https://www.ted.com/speakers/suzanne_simard.

Simila, Tiu, and Fernando Ugarte. "Surface and Underwater Observations of Cooperatively Feeding Killer Whales in Northern Norway." *Canadian Journal of Zoology* 71, no. 8 (1993): 1494–99.

Simmons, Charlotte. "Crisis in Our National Parks: How Tourists Are Loving Nature to Death." *Guardian,* November 20, 2018.

Sorenson, E. Richard. "Pre-Conquest Consciousness." In *Tribal Epistemologies: Essays in the Philosophy of Anthropology,* edited by Helmut Wautischer, 79–115. Burlington, VT: Ashgate, 1998.

Sortie, Bruce, Jeffrey Reimer, and Gordon Jones. *Oregon Agriculture, Food, and Fiber: An Economic Analysis.* College of Agricultural Sciences, Oregon State University, August 2021.

Souder, William. *On a Farther Shore: The Life and Legacy of Rachel Carson.* New York: Crown, 2012.

Speece, Darren. *Defending Giants: The Redwood Wars and the Transformation of American Environmental Politics.* Seattle: University of Washington Press, 2017.

Spence, Mark. 1999. *Dispossessing the Wilderness: Indian Removal and the Making of the National Parks.* New York: Oxford University Press, 1999.

Stegner, Wallace. Wilderness Letter. Written to the Outdoor Recreation Resources Review Commission, December 3, 1960. Published in "Wilderness Idea" in *The Sound of Mountain Water.* Garden City, NY: Doubleday, 1969.

Stone, Christopher. "Should Trees Have Standing? Towards Legal Rights for Natural Objects." *Southern California Law Review* 45 (1972): 450–501.

Tan, Zhai Yun. "Bees Awe Scientists by Displaying Learning and Teaching Skills." *Christian Science Monitor,* October 5, 2016.

Tattersall, Ian. *Masters of the Planet.* New York: Palgrave Macmillan, 2012.

Teixidor, Patricia. "Cultural Traditions in Orcas." *Mapping Ignorance,* February 25, 2013. mappingignorance.org/2013/02/25/cultural-traditions-in-orcas/#note.

Thompson, Nainoa. "On Wayfinding." https://archive.hokulea.com/ike/hookele/on_wayfinding.html.

Thoreau, Henry David. *The Indians of Thoreau: Selections from the Indian Notebooks.* Edited by Richard F. Fleck. Albuquerque, NM: Hummingbird Press, 1974.

Thoreau, Henry David. *The Journal, 1837–1861.* New York: New York Review Books, 2009.

Thoreau, Henry David. *Walden.* 1845. Gutenberg E-book #205.

"Thoreau's Views on Indians." Penobscot Cultural and Historic Preservation. https://www.penobscotculture.com/index.php/thoreau-s-views-on-indians.

Tremblay, Marie A., and Colleen C. St. Clair. "Permeability of a Heterogeneous Urban Landscape to the Movements of Forest Songbirds." *Journal of Applied Ecology* 48, no. 3 (June 2011): 679–88.

Tushingham, Shannon. *Archaeology, Ethnography, and Tolowa Heritage at Red Elderberry Place, Chvn-su'lh-dvn, Jedediah Smith Redwoods State Park.* California Department of Parks and Recreation Archaeology, History and Museums Division, 2013.

Tvesko, Mark. "Social Identity and Cultural Change on the Southern Northwest Coast." *American Anthropologist* 109, no. 3 (2007): 431–41.

United Nations Environment Program. *Mainstreaming Biodiversity in Production Landscapes.* 2018. https://wedocs.unep.org/handle/20.500.11822/26878.

U.S. Geologic Survey. "Moving Mountains: Elwha River Still Changing Five Years After World's Largest Dam-Removal Project." September 5, 2018.

U.S. National Marine Fisheries Service, Northwest Region. *Recovery Plan for Southern Resident Killer Whales (Orcinus orca).* National Oceanic and Atmospheric Administration. January 17, 2008. https://repository.library.noaa.gov/view/noaa/15975.

Utrilla, Pilar, Carlos Mazo, María Cruz Sopena, Manuel Bea, and Rafael Domingo Martínez. "A Paleolithic Map from 13,660 BP: Engraved Stone Blocks from the Late Magdalenian in Abauntz Cave." *Journal of Human Evolution* 57 (2009): 99–111.

Valdivia, Abel, Shaye Wolf, and Kieran Suckling. "Marine Mammals and Sea Turtles Listed Under the U.S. Endangered Species Act Are Recovering." *PLoS One* 14, no. 1 (January 16, 2019).

Vermeulen, Sonja J., Bruce M. Campbell, and John S. I. Ingram. "Climate Change and Food Systems." *Annual Review of Environment and Resources* 37 (November 2012): 195–222.

Von Petzinger, Genevieve. *The First Signs: Unlocking the Mysteries of the World's Oldest Symbols.* New York: Atria Books, 2016.

Walth, Brent. *Fire at Eden's Gate: Tom McCall and the Oregon Story.* Portland: Oregon Historical Society, 1994.

Watts, Alan. *Cloud-Hidden, Whereabouts Unknown: A Mountain Journal.* New York: Vintage, 1968.

Wegner, Dave. "Restoring Glen Canyon." Glen Canyon Institute. https://www.glencanyon.org/restoring-glen-canyon/.

Weitkamp, Laurie. "Hatchery and Wild Salmon in the Columbia River Estuary—Similar or Different?" Northwest Fisheries Science Center, Newport, OR, 2013.

Welch, Craig. "Groundbreaking Effort Launched to Decode Whale Language." *National Geographic,* April 2021.

Wells, Nancy, and Kristie Lekies. "Nature and the Life Course: Pathways from Childhood Nature Experiences to Adult Environmentalism." *Children, Youth, and Environments* 16, no. 1 (2006): 1–24.

Wells, Ray, David Bukry, Richard Friedman, Doug Pyle, Robert Duncan, Peter Haeussler, and Joe Wooden. "Geologic History of Siletzia, a Large Igneous

Province in the Oregon and Washington Coast Range: Correlation to the Geo-magnetic Polarity Time Scale and Implications for a Long-Lived Yellowstone Hotspot." *Geosphere* 10, no. 4 (2014): 692–719.

"Wes Jackson." *The Land Institute.* Landinstitute.org/about-us/staff/wes-jackson/.

White, Lynn Townsend, Jr. "The Historical Roots of Our Ecological Crisis." *Science* 155 (1967): 1203–7.

Whittle, Alastair. "'Very Like a Whale': Menhirs, Motifs, and Myths in the Mesolithic-Neolithic Transition of Northwest Europe." *Cambridge Archaeological Journal* 10, no. 2 (2000): 243–59.

Wilkinson, Richard G., and Kate Pickett. *The Spirit Level: Why Greater Equality Makes Societies Stronger.* New York: Bloomsbury Press, 2011.

Williams, Gerald. *References on the American Indian Use of Fire in Ecosystems.* Washington, DC: USDA Forest Service, 2001.

Wilson, Edward O. *The Social Conquest of the Earth.* New York: W. W. Norton, 2012.

Witter, Robert C., Eileen Hemphill-Haley, Roger Hart, and Lindsey Gay. *Tracking Prehistoric Cascadia Tsunami Deposits at Nestucca Bay, Oregon.* Final Technical Report to U.S. Geological Service, Oregon Department of Geology and Mineral Industries, Newport, OR, 2010.

Wohlleben, Peter. *The Hidden Life of Trees.* Vancouver, BC: Greystone Books, 2016.

Wood, Charlie. "How Smart Is a Bumblebee? Smarter Than You Think, Say Scientists." *Christian Science Monitor,* February 24, 2017.

Wright, Brianna M., Eva H. Stredulinsky, Graeme M. Ellis, and John K. B. Ford. "Kin-Directed Food Sharing Promotes Lifetime Natal Philopatry of Both Sexes in a Population of Fish-Eating Killer Whales, *Orcinus orca.*" *Animal Behaviour* 115 (May 2016): 81–95.

Xerces Society. "Bumble Bee Conservation." https://xerces.org/bumblebees.

Young, Christian C. *In the Absence of Predators: Conservation and Controversy on the Kaibab Plateau.* Lincoln: University of Nebraska Press, 2002. *Environment, Net Reviews,* December 2003.

Zeder, Melinda A. "Origins of Agriculture in the Near East." *Current Anthropology* 52, no. S4 (2011): 221–35.

Blome, in the Jordan and Washington Court Range: Correlation in the Great magnetic Volcanic Tuffs field and Implication. Jona Longathard Inowchane Blome." Chesapeare Co., No. 5 Long, 6024-030.

Wixz, Wilson. "The Coast Feature of authimentine organiloug-the railways are fauna. Wint, Lynn Townsend, Jr. "The Historical Echo of Out of Shopland cities," Art review, 1952, 1952-1152.

Whitby, Margaret. "Vere Like a Wharf Sea Form... Mikela," ... a Matter of the Marame World that Ten won of Northchaver Europe... Coarse view Anthropology and political magazine, 1992, 1955-55.

Williams, Richard C. and Kate Peters. The Wine Twins: Why Cooper Countries Matter. New women, New York, Bloombsbury Press, 2012.

Williams, Gerald. "My knowsen our American Team. Great Fire in Louisiana." Washington, D.C.: Nature Society, 2000.

Wilson, Edward O. The Social Conquest of the Earth, New York: W. W. Norton, 2012.

Wirtz, Robert I. "Lllece Hampfad Mine. Roger Harshald Joresy Cty. Bree Pty P. Blacom Carolina Aingman..." south of restrictard. I. Oregon Feedh Tabita cal Report no. 73. Geological Survey. Oregon Department of Geology and Mineral Industries. Notem OR, 2014.

Wohleforth, Tyre The Outline Life of Dale Vacuoing RE: the Snatch RR, 2010. Yogun Charlie. "How Snatch is a Bankrupted Interest. This Is Think good Sweet triles." One Lan Binet. Month of. February 1, 2017.

Wright, Barbara M., Eva H. Strudinsky, Corona, LR Rills, and John M. B. Load. "Kin Discovered Food sharing Promotes Unrelated Natal Philopatry of both Sexes in a Population of Adult Black Wardge Graveling. A broad Behaviour 73 (Mar 2016): 83-92.

Xerxes Society, "Bumble Bee Conservation." https://www.greenbattlebee... Young, Christian C. Defin. Manage of Predators: Population, Cycle of romance on the Kanah Platean. Hancoln: University of Nebraska Press, 2002. Enter review. Out Review. December 2006.

Zahn, Affahide A. "Scupture of Agriculture in the State Nature Coastal Bank and Raya Sacna. Sejzi Fallaza vaxes. Rawpyr...

Index

About the Author

MARION DRESNER earned a BA in biology from the University of New York, Buffalo, an MS in natural resources from Humboldt State University, and her PhD in natural resource management from the University of Michigan. Her early career focused on environmental education, and she worked as a national park ranger for four years. Dresner was a professor at Portland State University from 1995 until her retirement in 2018 and now lives in rural Oregon.